Revise AS Physics for AQA Specification A

Harvey Cole
and
David Sang

Contents

Introduction – How to use this revision guide

This revision guide is designed to follow the **AQA Advanced Subsidiary GCE in Physics** (Specification A – 5451) and is divided into three modules to match the course. You may be taking a test at the end of each module, or you may take all the tests at the end of the course. The content is exactly the same.

Each module begins with an **introduction**, which summarises the content. It also reminds you of the topics from your GCSE course which the module draws on.

The content of each module is presented in **blocks**, to help you divide up your study into manageable chunks. Each block is dealt with in several double-page spreads. These do the following:

- they **summarise** the content;
- they indicate **points to note**;
- they include **worked examples** of calculations;
- they include **diagrams** of the sort you might need to reproduce in tests;
- they provide **hint boxes** to give you general hints about how to get the Physics right;
- they provide **quick check questions** to help you test your understanding;
- they provide **exam-style questions** at the end, similar to those you will encounter in exams.

To show that you are heading in the right direction, **answers** to all questions are provided at the end of the book.

You need to understand the **scheme of assessment** for your course. This is summarised on page iv overleaf.

A note about units

In the worked examples, we have included units throughout the calculations. This can help to ensure that you end up with the correct units in your final answer. See also the note on checking units on page 35.

In an examination it is not necessary to include the units in all steps of a calculation. Always include units in the final answer.

ROCHDALE & MIDDLETON

AQA AS Physics – Assessment

There are three **units of assessment** (Units 1, 2 and 3) in this AS Physics course, one for each module. Unit 3 includes assessment of **experimental skills.**

Unit	Name	Duration of written test	Types of question	Weighting (% of AS marks)
1	Particles, radiation and quantum phenomena	90 minutes	short structured (60 marks)	30%
2	Mechanics and molecular kinetic theory	90 minutes	short structured (60 marks)	30%
3	Current electricity and elastic properties of solids	75 minutes	short structured (50 marks)	25%
+ *either*	Experimental skills	varies	coursework (30 marks)	15%
or	Experimental skills	90 minutes	practical (30 marks)	15%

Short structured questions require brief answers to several linked parts of a question.

Your answers will be used to assess the quality of your **written communication.** Up to two marks in each paper will be awarded for your ability to:

- select an appropriate style of writing,
- organise the information,
- use specialist vocabulary,
- write legibly, using accurate spelling, grammar and punctuation.

> In exams, use the mark allocation and the space available for your answer to guide how much you write.

Module 1: Particles, radiation and quantum phenomena

To help you organise your learning, each module is broken down into blocks. There are three blocks in this module.

- **Block 1A** is the longest. It is about the particles that make up an atom, and introduces the concept of antiparticles. Much of this is 'new' work and does not draw on GCSE topics.
- **Block 1B** shows how reflection and refraction of light can be represented by rays. Snell's law shows how to predict the path of a refracted ray.
- **Block 1C** introduces some ideas about the fundamental nature of matter and radiation. We are used to thinking of matter as being made up of particles, and of radiation as waves. Here you will learn that things are not so simple. Sometimes matter behaves as a wave, and radiation may behave as particles.

You will need to be able to deal with large powers of 10, so it is important you learn how to operate your calculator quickly and correctly!

Block 1A: Particles and antiparticles, pages 2–9

Ideas from GCSE	Content outline of Block 1A
● Existence and structure of atoms	● Structure of atoms ● Evidence for the existence of the nucleus ● Particles, antiparticles, photons ● Classifying particles ● Quarks and antiquarks

Block 1B: Refraction, pages 10–13

Ideas from GCSE	Content outline of Block 1B
● Reflection and refraction of light ● Total internal reflection and its applications	● Laws of reflection ● Why refraction happens ● Laws of refraction, including Snell's law ● Refractive index, n ● Total internal reflection and critical angle, θ_c ● Modern applications of fibre optics

Block 1C: Electromagnetic radiation and quantum phenomena, pages 14–19

Ideas from GCSE	Content outline of Block 1C
● The electromagnetic spectrum ● Speed, frequency and wavelength of a wave	● Photons and photon energy ● The photoelectric effect ● Line spectra and energy levels in atoms ● Wave–particle duality

End-of-module questions, pages 20–22

Nuclear structure

Atoms are made of **protons**, **neutrons** and **electrons**. The protons and neutrons make up the **nucleus**, with the electrons orbiting around it.

Particles in an atom

particle	mass	charge
proton	1	+1 (positive)
neutron	1	0 (uncharged)
electron	1/1840	−1 (negative)

Protons and neutrons are **nucleons** (particles found in the nucleus).

Masses are given relative to the proton. The mass of a neutron is very slightly more than that of a proton.

Charges are in units of $e = 1.6 \times 10^{-19}$ C.

Representing a nucleus

Z = proton number (or **atomic number**) = number of protons in nucleus

N = neutron number = number of neutrons in nucleus

A = nucleon number (or **mass number**) = number of nucleons in nucleus

Since protons and neutrons are nucleons, $A = Z + N$.

In a neutral atom, number of electrons = number of protons = Z.

An individual nucleus can be represented thus: $_{Z}^{A}X$ e.g. $_{8}^{16}O$ and $_{92}^{238}U$

Each combination of A and Z represents a different nuclear species or **nuclide**.

Isotopes

Each element exists in a variety of forms or **isotopes**. Each isotope has the same number of protons in the nucleus, but different numbers of neutrons, so their masses are different.

Alpha-particle scattering

An atom is normally neutral – it contains equal amounts of positive and negative charge. The positive charge is concentrated in the tiny nucleus at the centre. The negative charge (of the electrons) is spread out around the nucleus. Evidence for this comes from Rutherford's **alpha-particle scattering experiment**.

✓ **Quick check 1, 2**

$_{1}^{1}H \quad _{1}^{2}H$

two isotopes of hydrogen

$_{26}^{54}Fe \quad _{26}^{56}Fe$

two isotopes of iron

✓ **Quick check 3**

gold foil

alpha-particles

back-scattered alpha-particle

alpha-particles

nucleus

- A narrow beam of alpha-particles was aimed at a thin gold foil.
- An alpha-particle is made of 2 protons and 2 neutrons, so has a positive charge.
- The direction of the alpha-particles after they had passed through the foil was determined using a detector.
- Most of the alpha-particles went straight through.
- A few were deflected more than 90°.

evidence	deduction
Most alpha-particles pass straight through gold foil.	Atoms are mostly 'empty space'.
A few (1 in about 10^4) are deflected back towards the observer.	Positive charge is concentrated in a tiny volume – the nucleus.

Quick check 4

Relative sizes

diameter of nucleus $\sim 10^{-15}$ m (1 fm)

diameter of atom $\sim 10^{-10}$ m (0.1 nm)

A **molecule** may be similar in size to a single atom, or it may be much larger, depending on how many atoms it is made of. A protein molecule may have a diameter of 10^{-7} m.

Quick check 5, 6

Quick check questions

1. Represent in symbolic form a nucleus of a silicon (Si) atom that consists of 14 protons and 14 neutrons.

2. How many protons, neutrons and electrons are there in a neutral oxygen atom whose nucleus is represented by $^{16}_{8}O$?

3. The table shows the composition of four nuclei. Which nuclei are isotopes of the same element?

nucleus	number of protons	number of neutrons
A	24	26
B	23	26
C	23	27
D	24	27

4. In the alpha-scattering experiment, Rutherford used gold foils of different thicknesses. Explain how increasing the thickness of the foil would change the number of alpha-particles back-scattered.

5. By how many orders of magnitude (factors of 10) does the diameter of an atom exceed that of its nucleus?

6. The smallest object resolvable using an optical microscope has a diameter of the order of 1 μm. Roughly how many atomic diameters is this?

Classifying particles

Atoms are made of protons, neutrons and electrons. However, there are many other particles in nature and it is important to understand how these can be organised systematically.

Particles and antiparticles

For each particle of **matter**, there is an **antiparticle**. Antiparticles make up **antimatter**. Antiparticles are usually represented by putting a bar over the symbol for the particle (e.g. $\bar{p}$ for antiproton).

particle	antiparticle	nature of antiparticle
electron, e or e^-	positron, $\bar{e}$ or e^+	same mass as electron; same charge but positive
proton, p	antiproton, $\bar{p}$	same mass as proton; same charge but negative
neutrino, ν	antineutrino, $\bar{\nu}$	same mass as neutrino; zero charge; spins in opposite direction to neutrino

✓ **Quick check 1**

Photons of electromagnetic radiation

Sometimes electromagnetic radiation (such as light) behaves as if it were made of particles rather than waves. These particles are called **photons**. Photon energy E is related to frequency f by

$$E = hf$$

where h is the Planck constant: $h = 6.63 \times 10^{-34}$ J s.

High-frequency radiation (such as X-rays and gamma rays) has the most energetic photons. Low-frequency radiation (such as radio waves and microwaves) has less energetic photons.

There is no such thing as an anti-photon; antimatter emits photons identical to those emitted by 'normal' matter. The symbol for a photon is γ (gamma) – even if it isn't a gamma-photon.

✓ **Quick check 2**

Annihilation and creation

When a particle meets its antiparticle, they may annihilate one another. A photon of electromagnetic energy is produced; this is **annihilation**.

For example, a proton and an antiproton annihilate: $p + \bar{p} \rightarrow \gamma$.

The same process can operate in reverse: if a photon has enough energy, a particle–antiparticle pair can be produced; this is **pair production**.

For example, a photon gives rise to an electron–positron pair: $\gamma \rightarrow e + \bar{e}$.

✓ **Quick check 3**

Leptons, mesons and baryons

Studies of the interactions of the many types of particles have been explained by classifying particles according to the following scheme. The heaviest particles are on the right. There are two 'families': **leptons** ('light ones') and **hadrons** ('heavy ones'). Hadrons are divided into two types according to how many **quarks** they are made of.

leptons	hadrons	
	mesons	baryons
e.g. electron, muon, neutrino	e.g. pion, kaon	e.g. proton, neutron
thought to be indivisible (i.e. fundamental)	made of 2 quarks	made of 3 quarks

✓ *Quick check 4*

Quarks and antiquarks

Isolated quarks have not yet been observed; they are always grouped together in twos and threes as hadrons. Quarks, like leptons, are believed to be fundamental (indivisible) particles of matter. They have odd-sounding names, such as up, down and strange, abbreviated to u, d and s. For each quark, there is a corresponding **antiquark**.

- A **meson** is made from a quark and an antiquark.
- A **baryon** is made from 3 quarks; change all the quarks into antiquarks to make the antiparticle (e.g. proton = u u d, antiproton = $\bar{u}\ \bar{u}\ \bar{d}$).

▶▶ *More about quarks on pages 6–7.*

mesons

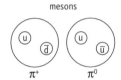

π^+ π^0

baryons

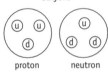

proton   neutron

✓ *Quick check 5*

> ❓ *Quick check questions*
>
> 1 The charge on an electron is -1.6×10^{-19} C. What is the charge on a positron?
>
> 2 Calculate the energy of a photon of gamma radiation of frequency 5×10^{22} Hz. (The Planck constant, $h = 6.63 \times 10^{-34}$ J s.)
>
> 3 Write an equation to represent the annihilation of a positron and an electron when they meet.
>
> 4 A φ particle is made of 2 quarks. Which of the following categories does it belong to: lepton, hadron, meson, baryon?
>
> 5 A neutron is made from the following 3 quarks: u d d. What is an antineutron made from?

Don't worry about the odd-sounding names given to quarks and their properties. Simply learn to use them like a kind of alphabet from which particles are made.

Interactions between particles

Particle physics seeks to explain what happens when particles interact. An individual particle may decay into other particles. Two or more particles may collide. What forces are at work?

Quarks and hadrons

To understand interactions involving hadrons, it is necessary to know which quarks each hadron is made of.

hadron	quarks	hadron	quarks	hadron	quarks
proton, p	u u d	pion, π^+	u $\bar{\text{d}}$	kaon, K^+	u $\bar{\text{s}}$
neutron, n	u d d	pion, π^-	$\bar{\text{u}}$ d	kaon, K^-	$\bar{\text{u}}$ s
		pion, π^0	u $\bar{\text{u}}$ *or* d $\bar{\text{d}}$		

✓ Quick check 1, 2

Stability and decay

Some radioactive nuclei decay by emitting a beta particle (an electron). An antineutrino is also emitted. Inside the nucleus, a neutron has decayed to become a proton:

$$n \rightarrow p + e^- + \bar{\nu}$$

By comparing the quark composition of a neutron (u d d) and a proton (u u d), we can see that a down quark (d) has turned into an up quark (u):

$$d \rightarrow u + e^- + \bar{\nu}$$

This is known as β^- **decay**, because the particle emitted is a negatively charged electron. Some nuclei decay by emitting a positron (positive charge). This is known as β^+ **decay**.

$$p \rightarrow n + e^+ + \nu$$
$$u \rightarrow d + e^+ + \nu$$

The proton is the most stable and long-lived of all hadrons. An isolated neutron (outside a nucleus) soon decays. Other hadrons decay even more rapidly, which is why they are rarely observed except in high-energy experiments.

✓ Quick check 3, 4

Conservation laws

In any interaction between particles, several properties are conserved. For example, when a neutron decays, charge is conserved:

$$n \rightarrow p + e^- + \bar{\nu} \qquad 0 \rightarrow (+1) + (-1) + 0$$

Other conserved properties include **baryon number** and **strangeness**.

quark	charge Q	baryon number B	strangeness S
up, u	$+\frac{2}{3}$	$\frac{1}{3}$	0
down, d	$-\frac{1}{3}$	$\frac{1}{3}$	0
strange, s	$-\frac{1}{3}$	$\frac{1}{3}$	-1
For antiquarks, reverse the signs of *all* these quantities.			

> ▶ Don't think too hard about what these properties mean. You just need to be able to check that each is conserved in any interaction involving quarks.

Worked example

Show that charge, baryon number and strangeness are conserved in the following interaction:

$$p + p \rightarrow p + p + \pi^0$$

Step 1   Replace hadron symbols by their quarks:

$$uud + uud \rightarrow uud + uud + u\bar{u}$$

Step 2   Check each quantity in turn:

$$\text{Charge on left} = \left(\frac{2}{3} + \frac{2}{3} - \frac{1}{3}\right) + \left(\frac{2}{3} + \frac{2}{3} - \frac{1}{3}\right) = +2$$

$$\text{Charge on right} = \left(\frac{2}{3} + \frac{2}{3} - \frac{1}{3}\right) + \left(\frac{2}{3} + \frac{2}{3} - \frac{1}{3}\right) + \left(\frac{2}{3} - \frac{2}{3}\right) = +2$$

Hence charge is conserved.

$$\text{Baryon number on left} = \left(\frac{1}{3} + \frac{1}{3} + \frac{1}{3}\right) + \left(\frac{1}{3} + \frac{1}{3} + \frac{1}{3}\right) = +2$$

$$\text{Baryon number on right} = \left(\frac{1}{3} + \frac{1}{3} + \frac{1}{3}\right) + \left(\frac{1}{3} + \frac{1}{3} + \frac{1}{3}\right) + \left(\frac{1}{3} - \frac{1}{3}\right) = +2$$

Hence baryon number is conserved.

For all of these quarks, strangeness $S = 0$ so strangeness is also conserved.

✓ *Quick check 5*

❓ Quick check questions

1 Write down the quark composition of an antiproton.

2 Explain why an anti-π^0 is the same as a π^0.

3 Which particles are formed when a neutron decays? And when an antineutron decays?

4 Draw a diagram to show the change that occurs to the quarks of a proton during β^+ decay.

5 Which of the following quantities is *not* conserved in the following interaction: charge, baryon number and strangeness?

$$p + p \rightarrow p + p + \pi^+ + \pi^0$$

Representing forces

To explain the interaction between two particles, we use the idea of **exchange particles**. To show what is happening during an interaction, we draw a **Feynman diagram**.

Exchange particles

Two electrons repel one another because they both have negative electric charge ('like charges repel'). We picture this force being transmitted by a photon passing between them. The photon is the exchange particle that transmits the force.

For each fundamental force in nature, there is a corresponding exchange particle. For simplicity, we will consider only two forces:

force	example	exchange particles	symbol
electromagnetic force	attraction and repulsion of charges	photons	γ
weak nuclear force	involved in beta decay	intermediate vector bosons	W^+, W^-, Z

Intermediate vector bosons can be detected in particle accelerators. They disappear; particles and antiparticles appear. For example, $Z \rightarrow e^- + e^+$.

▶▶ *Compare this with pair production when a photon disappears – page 4.*

✓ Quick check 1

Feynman diagrams

These diagrams represent interactions between particles. Things to look for:

- The exchange particles are shown by wavy lines.
- The particles themselves are shown by straight lines.
- Look for arrows going in and arrows coming out, to follow the sequence of events.
- Check that charge is conserved at each vertex of the diagram.

Example 1: Repulsion between electrons

This is straightforward. The two electrons approach one another, exchange a photon, and move apart. This is shown in the top diagram.

Example 2: β^- decay of a neutron

The neutron emits a W^- boson and becomes a proton. Then the W^- decays to become an electron and an antineutrino. See the middle diagram.

In terms of quarks, one of the down quarks in the neutron emits a W^- boson and becomes an up quark. This is shown in the lower diagram.

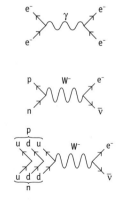

✓ Quick check 2

More Feynman diagrams

An antineutrino collides with a proton to form a neutron and a positron. This reaction is used to detect neutrinos.

✓ Quick check 3

A proton in the nucleus of an atom captures an orbiting electron. This is called **electron capture**.

✓ Quick check 4, 5

? *Quick check questions*

1 Which type of exchange particle is involved when a neutron decays? And when two protons repel one another?

2 Draw a Feynman diagram to show the decay of a down quark to become an up quark. Write a short description of the process.

3 Write an equation to represent an antineutrino–proton collision. Draw a Feynman diagram to represent this as a quark–antineutrino interaction.

4 Write an equation to represent electron capture.

5 Write a brief description of the interaction represented by this Feynman diagram. Include an equation.

▶ Represent the proton and neutron by their quarks.

Reflection and refraction

To explain how light behaves, we can think of light travelling as **rays**. A ray travels in a straight line. It will change direction if:
- it is **reflected** (when it strikes a surface);
- it is **refracted** (when it passes from one material to another).

▶▶ *We can also think of light as waves – see pages 18–19.*

Laws of reflection

The **laws of reflection** tell you where a ray will go when it is reflected. The **normal** is the line at 90° to the reflecting surface at the point where the incident ray strikes it.
- **Law 1:** Angle of incidence = angle of reflection, $i = r$ (angles measured from normal to ray).
- **Law 2:** Incident ray, reflected ray and normal are all in the same plane.

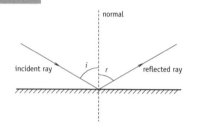

✓ *Quick check 1*

Refraction: when it happens

Light travels fastest in a vacuum. It travels more slowly in other media. When light changes speed (because it travels from one medium to another), it is *refracted*.
- If a ray enters a medium head-on (angle of incidence $i = 0$), it travels straight on.
- If a ray enters a medium obliquely, it changes direction.

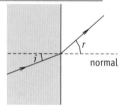

light speeding up: ray bends away from normal

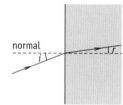

light slowing down: ray bends towards normal

Laws of refraction

As with reflection, angles are measured from the normal to the ray.

Law 1: Incident ray, refracted ray and normal are all in the same plane.

Law 2: Snell's law (see opposite) explains how the angles of incidence and refraction are related.

▶ The angle r is now the angle of refraction, not reflection.

✓ *Quick check 2*

Refractive index *n*

The **refractive index *n*** of a medium relates the speed of light in the medium to the speed of light in free space (a vacuum).

$$\text{refractive index } n = \frac{\text{speed of light in free space}}{\text{speed of light in medium}} = \frac{c_o}{c_{medium}}$$

In a medium of refractive index 2.0, light travels at half its speed in free space. Some values are worth remembering:
- $n_0 = 1$ (by definition)
- $n_{air} = 1.00$ (to 2 decimal places)
- $n_{water} = 1.33$
- $n_{glass} \sim 1.5$ (depending on the composition of the glass)

✓ *Quick check 3*

Snell's law

For a ray passing from air into a medium of refractive index n, the angle of incidence i and the angle of refraction r are related by:

$$n = \frac{\sin i}{\sin r}$$

Where a ray passes from medium 1 into medium 2 (speeds c_1 and c_2) at angles θ_1 and θ_2 to the normal, calculate the relative refractive index $_1n_2$ using:

$$_1n_2 = \frac{c_1}{c_2} = \frac{\sin \theta_1}{\sin \theta_2}$$

If the refractive index for medium 1 is n_1 and that for medium 2 is n_2:

$$_1n_2 = \frac{n_2}{n_1}$$

Worked example

A ray of light travels from glass ($n_g = 1.5$) into water ($n_w = 1.33$) with an angle of incidence θ_g of 30° – see diagram. Calculate the angle of refraction θ_w.

Step 1 Calculate the relative refractive index from the values for the two materials:

$$_gn_w = \frac{n_w}{n_g} = \frac{1.33}{1.5} = 0.887$$

Step 2 Substitute values into the Snell's law equation, rearrange and solve:

$$_gn_w = \frac{\sin \theta_g}{\sin \theta_w}$$

$$0.887 = \frac{\sin 30°}{\sin \theta_w}$$

$$\sin \theta_w = \frac{\sin 30°}{0.887} = 0.564$$

$$\theta_w = 34°$$

The refractive index doesn't change much, so the change of direction is small.

$_gn_w$ is less than 1 because the light speeds up as it enters the water.

✓ *Quick check 4*

? Quick check questions

1 What are the values of the angles of incidence and reflection (upper diagram)?

2 Does the ray speed up or slow down when it enters medium 2 (lower diagram)?

3 Which medium in question 2 has the higher refractive index?

4 A ray of light, travelling through air, strikes a glass surface with an angle of incidence of 40°. The refractive index of the glass is 1.47. Draw a diagram to show the situation. Calculate the angle of refraction.

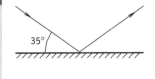

Total internal reflection

Total internal reflection (TIR) of light may occur when a ray is travelling inside a glass block. The ray reaches the edge of the block; what happens next depends on the angle of incidence i.

Note that there is always a weaker reflected ray as well.

- **Total** – because 100% of the light is reflected.
- **Internal** – because the ray is reflected *inside* the material.
- **Reflection** – because the light is reflected, not refracted.

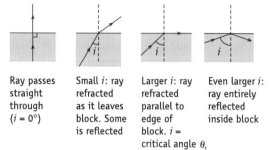

Ray passes straight through ($i = 0°$)

Small i: ray refracted as it leaves block. Some is reflected

Larger i: ray refracted parallel to edge of block. $i =$ critical angle θ_c

Even larger i: ray entirely reflected inside block

Critical angle θ_c

Total internal reflection can only happen when a ray travelling through a material of *higher* refractive index reaches the boundary with a material of *lower* refractive index.

Total internal reflection happens for any angle of incidence equal to or greater than the **critical angle** θ_c.

At the critical angle, $i = \theta_c$ and $r = 90°$. Hence $\sin i = \sin \theta_c$ and $\sin r = 1$. Snell's law then gives:

$$n = \frac{1}{\sin \theta_c} \quad \text{or} \quad \sin \theta_c = \frac{1}{n}$$

For glass of refractive index 1.5, $\sin \theta_c = 1/1.5 = 0.667$ and $\theta_c = 42°$, approximately.

✓ *Quick check 1, 2*

Using TIR in optic fibres

A ray of light can travel along inside a solid glass fibre. Each time it reaches the outer surface of the glass it is reflected back inside, since i is nearly 90°.

Practical optic fibres

Optic fibres are made from glass or plastic, surrounded by a **cladding** of material with a slightly lower refractive index.

Rays that travel straight down the centre of the fibre have the shortest route and take least time. Oblique rays have further to travel, and take longer.

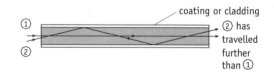

coating or cladding

② has travelled further than ①

✓ *Quick check 3*

Transmitting data

Optic fibres carry data in digital form. A ray of light from a laser is modulated (switched on and off) at high frequency to encode the data, rather like Morse code. Problems arise if rays can travel along different paths inside the fibre.

a single pulse enters the fibre

the pulse is 'smeared': some rays have taken longer than others

This smearing of a pulse is called **multipath dispersion**. To avoid this problem, most fibres are made with a very narrow core so that all rays pass virtually straight down the middle.

✓ *Quick check 4*

Advantages of optic fibres

Optic fibres are used for **endoscopy.**

An **endoscope** is a flexible tube containing optic fibres that is inserted into a patient's body to inspect internal surfaces without the need for surgery. There are two fibre optic bundles – one to send light to the area under investigation, the other to transmit an image back to the observer.

Optic fibres have also made possible the **Internet.**

They are used for:

- telecommunications networks (carrying telephone messages),
- cable television,
- links between computers (for high-speed data transfer).

Digital signals are less susceptible to noise than analogue signals. Because of the high frequency of light, optic fibres can carry vastly more data than an electric current in a cable of comparable size. They are very difficult to bug. Because the light is only weakly absorbed, signals can travel many kilometres before they become so weak that they need to be regenerated.

Quick check questions

1 Calculate the critical angle for glass of refractive index 1.6.
2 Calculate the critical angle at the interface between glass ($n = 1.6$) and water ($n = 1.33$).
3 Explain why the central core of an optic fibre must be coated with a material of *lower* refractive index.
4 Different wavelengths of light travel at different speeds through glass. Explain why white light could not be used for long-distance information transfer. Why is laser light suitable?

The photoelectric effect

When light shines on certain metals, electrons break free. This is the **photoelectric effect.**

In order to explain this effect, Albert Einstein had to assume that, when light interacts with a metal, it behaves as particles (photons), not as waves. The energy of a photon of light is captured by a *conduction electron* in the metal, and the electron escapes from the surface of the metal.

Observing the photoelectric effect

When light is shone onto the photocell, a current starts to flow immediately. Even a very feeble light will work. The brighter the light, the greater the current.

The explanation is that the energy of the light helps conduction electrons to break free from the metal cathode. They cross to the anode; now there is a flow of charge all round the circuit. More light means more energy in terms of photons per second, so more electrons break free every second.

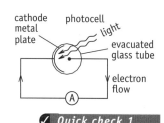

✓ *Quick check 1*

Explaining the photoelectric effect

- Electrons don't normally escape from a metal.
- The conduction electrons are weakly held inside the metal.
- They need some energy to escape.
- Light can provide the necessary energy.

Why waves can't explain the effect

If we picture light waves falling on the metal, their energy is spread out all over the surface of the metal. It would take a long time for enough energy to be captured by the metal to free any electrons. The photoelectric effect is surprising because electrons break free as soon as the light is switched on.

Einstein argued that the energy of the light must be concentrated in tiny packets (photons). An individual conduction electron in the metal captures an individual photon; now the electron has enough energy to escape from the metal.

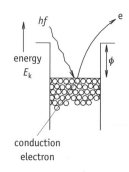

In the figure,

- hf is the energy of the photon,
- E_k is the kinetic energy of the electron,
- ϕ is the work function of the metal – the least amount of energy needed for an electron to escape from the surface of the metal.

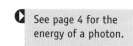

See page 4 for the energy of a photon.

How photons explain the effect

If a photon is captured by an electron in the metal, some of its energy is used to overcome the work function, and the rest ends up as the electron's kinetic energy.

The electrons highest up in the energy 'well' are the most energetic electrons. When one of these electrons captures a photon and escapes, it will have the maximum possible kinetic energy, $E_{k\,max}$. This gives us the **'photoelectric' equation**:

$$hf = \phi + E_{k\,max}$$

Brighter light means more photons, so more electrons released. However, this doesn't increase the kinetic energy of the fastest electrons, because the individual photons do not have more energy. The current produced is proportional to the intensity of the light: greater intensity means more photons per second, so more electrons are released per second.

Threshold frequency

The frequency of the light must be above a certain minimum value, the **threshold frequency** $f_{threshold}$. Below this value, an individual photon does not have enough energy for an electron to overcome the work function. Hence:

$$hf_{threshold} = \phi$$

✓ *Quick check 2*

The photoelectric equation may be represented as a graph of $E_{k\,max}$ against f.

- The negative intercept on the $E_{k\,max}$ axis gives the work function ϕ.
- The slope gives the Planck constant h.
- The intercept on the f axis gives the threshold frequency $f_{threshold}$.

Worked example

A metal has a work function $\phi = 1.3 \times 10^{-19}$ J. What is the greatest possible kinetic energy of an electron released by a photon of energy 2.4×10^{-19} J?

Of the 2.4×10^{-19} J of energy provided by the photon, 1.3×10^{-19} J is used up in overcoming the work function. This leaves 1.1×10^{-19} J of kinetic energy.

$$E_{k\,max} = (2.4 \times 10^{-19}\,\text{J}) - (1.3 \times 10^{-19}\,\text{J}) = 1.1 \times 10^{-19}\,\text{J}$$

You might find it easier to work in electronvolts (eV) rather than J. See page 16 for the definition of the electronvolt as a unit of energy. Don't mix the two units!

✓ *Quick check 3*

? *Quick check questions*

Planck constant $h = 6.63 \times 10^{-34}$ J s; electron charge $e = 1.6 \times 10^{-19}$ C.

1 A photocell makes a current flow around a circuit. Does the current *inside* the photocell flow from + to −, or from − to +?

2 What is the threshold frequency for a metal surface whose work function is 2.4×10^{-19} J?

3 Light of frequency 3.0×10^{14} Hz falls on a metal surface whose work function is 1.6×10^{-19} J. Calculate the kinetic energy of the fastest electrons produced.

Line spectra, energy levels and ionisation

Some spectra appear as line spectra. Each line has a definite colour or wavelength. This tells us two things: electrons in atoms can have only certain values of energy (they occupy **energy levels** in atoms), and light is emitted as 'particles' called **photons**.

The electronvolt – an energy unit

Photon energies are very small. The **electronvolt (eV)** is a more convenient unit than the joule. 1 eV is the energy transferred when an electron moves between two points separated by a p.d. of 1 V.

$$1 \text{ eV} = 1.6 \times 10^{-19} \text{ J}$$

- To convert from J to eV: divide by 1.6×10^{-19} (i.e. multiply by 6.25×10^{18}).
- To convert from eV to J: multiply by 1.6×10^{-19}.

> Remember the equation $W = QV$ (energy = charge × voltage). That's where the definition of the electronvolt comes from. The electron charge is $e = 1.6 \times 10^{-19}$ C.

> ✓ **Quick check 1, 2**

Energy levels

The electrons in an isolated atom can only have certain fixed amounts of energy. These are called the **energy levels** of the atom.

An electron will normally occupy the lowest available energy level, called its **ground state**. If it receives exactly the right amount of energy, an electron can move to a higher level and the atom is said to be **excited**. It may fall back to a lower energy level and lose energy. This energy is emitted as a single photon of light (electromagnetic radiation) of a particular frequency.

Energy levels are usually given in eV. This is more convenient than joules.

When an electron moves from level E_1 to level E_2 a photon of frequency f is emitted:

$$hf = E_1 - E_2$$

where h = the Planck constant = 6.63×10^{-34} J s.

This links a particle property (photon energy) to a wave property (frequency).

▶▶ More about particles and waves on pages 18–19.

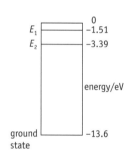

> Each line has negative energy, but this is a convention – it's the **difference** in levels that matters.

Worked example

Calculate the frequency of the radiation emitted for an energy transition in a hydrogen atom from level E_1 (−1.51 eV) to level E_2 (−3.39 eV).

Step 1 Photon energy = $E_1 - E_2$ = (−1.51) − (−3.39) = 1.88 eV.

Step 2 Change eV to J: 1.88 eV = $1.88 \times 1.6 \times 10^{-19}$ = 3.01×10^{-19} J.

Step 3 $f = (E_1 - E_2)/h = 3.01 \times 10^{-19}/6.63 \times 10^{-34} = 4.54 \times 10^{14}$ Hz.

> If you need the wavelength, use $c = f\lambda$ where c = speed of electromagnetic radiation = 3.0×10^8 m s^{-1}.

> ✓ **Quick check 3, 4**

Line spectra

The atoms of a gas at low pressure may be excited by a flame or by high-speed electrons (such as in a sodium vapour lamp). The excited gas atoms fall back to their ground state and emit photons of distinct energies corresponding to the differences between energy levels.

If you look at the flame or lamp through a diffraction grating, you will see lines of different colours. This is a **line spectrum** and is evidence for the existence of energy levels.

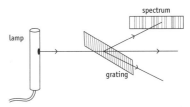

The atoms of any given element all have the same set of energy levels. This means that each element produces a unique line spectrum that may be used to identify the element.

Ionisation

Ionisation means giving an atom enough energy that an electron is completely removed from the atom. This is called the **ionisation energy** and can be provided by heating, or by collision with a high-speed electron, or by the atom 'capturing' a photon of sufficient energy.

To remove an electron completely from an atom requires energy to overcome the force of attraction between the negative charge on the electron and the positive charge on the nucleus.

The ionisation energy of hydrogen is −13.6 eV. A free electron would have to move through a potential difference or voltage of 13.6 V to gain enough kinetic energy (i.e. 13.6 eV) to ionise a hydrogen atom by colliding with it.

✓ *Quick check 5*

Fluorescent lamps

A fluorescent light tube contains mercury vapour gas at low pressure, and two electrodes. A large number of electrons are released from the cathode (negative electrode) and accelerated by the potential difference between anode (positive electrode) and cathode. Some of these electrons ionise gas atoms by collision, producing more electrons.

Some of the electrons collide with gas atoms which become excited and emit photons of ultraviolet light as they fall back to the ground state. These photons in turn excite atoms of a coating material on the inside of the tube, which emit visible light. Because the coating atoms are close together, there are so many 'allowed' energy transitions that all colours of the visible spectrum are produced.

? Quick check questions

Planck constant $h = 6.63 \times 10^{-34}$ J s; electron charge $e = 1.6 \times 10^{-19}$ C.

1 How many electronvolts of energy are transformed when an electron moves through a p.d. of 10.5 V?

2 What is the energy in eV of each photon in a beam of light of frequency 5×10^{14} Hz?

3 Four of the energy levels of an element are E_1 (ground state), E_2, E_3 and E_4. Transitions from any one level to any other will produce a photon. How many different photon wavelengths can be produced from the four levels?

4 An electron falls from an energy level $E_1 = -2.3 \times 10^{-19}$ J to a lower level $E_2 = -3.8 \times 10^{-19}$ J. Calculate the frequency of the light emitted.

5 The ionisation energy of argon is 2.5×10^{-18} J. State the minimum voltage needed to accelerate an electron so that it has enough kinetic energy to ionise an argon atom.

▶ First change joules to electronvolts.

Wave–particle duality

For Einstein to explain the photoelectric effect, he had to assume that, when light interacts with the conduction electrons in a metal, it behaves as particles (photons). At other times, we know that light behaves as waves – for example, when it is diffracted (spread out) as it passes through a slit, or when two light waves interfere with one another. So light can behave as waves or as particles, depending on the circumstances. This is known as **wave–particle duality**.

In a similar way, particles (such as electrons) may also show wave-like behaviour.

Electron diffraction

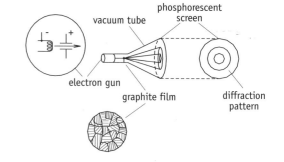

- Electrons can be diffracted – this shows that, when they pass through a fine grid, they behave like waves.

- A beam of fast-moving electrons is produced in a cathode ray tube.

- The electron beam passes through a thin layer of crystalline graphite (carbon).

- A diffraction pattern of fuzzy, light-and-dark rings is produced on a screen.

To make the electrons go faster, increase the accelerating voltage. The diameter of the rings decreases. This shows that the wavelength decreases as the electrons go faster.

✓ **Quick check 1, 2**

You can imagine how, as the wavelength of the waves gets smaller and smaller, the waves pass more easily through the gaps between the layers of carbon atoms, so they are diffracted less.

The de Broglie equation

We have already seen one equation that links a wave property (frequency f) with a particle property (photon energy E):

$$E = hf$$

The **de Broglie equation** is another equation that allows us to translate between wave behaviour and particle behaviour. It links wavelength λ with particle momentum p. The more momentum a particle has, the shorter its wavelength:

$$\lambda = \frac{h}{p}$$

Note that, in both of these equations, the Planck constant h connects the wave quantity to the particle quantity.

A note on momentum

You can calculate your momentum by multiplying your mass by your velocity, $m \times v$. The greater the mass of a particle, and the faster it moves, the greater its momentum. The formula $m \times v$ doesn't apply to fast-moving particles with speeds approaching the speed of light, so it is better to use the symbol p for momentum. The units of momentum are $kg\ m\ s^{-1}$ or N s.

▶▶ *More about momentum in Module 2.*

Worked example

An electron has momentum $2.00 \times 10^{-24}\ kg\ m\ s^{-1}$. What is its wavelength?

$$\lambda = \frac{h}{p} = \frac{6.63 \times 10^{-34}\ \text{J s}}{2.00 \times 10^{-24}\ \text{kg m s}^{-1}} = 3.31 \times 10^{-10}\ \text{m}$$

(This value of momentum is for an electron moving at roughly $2 \times 10^{6}\ m\ s^{-1}$. Its wavelength is similar to the size of an atom, which is why such electrons are diffracted by crystalline graphite.)

✓ *Quick check 3,4*

The de Broglie equation applies to *all* particles, no matter how big. If you run, you are like a moving particle. Your momentum might be $500\ kg\ m\ s^{-1}$; your wavelength would then be about $10^{-32}\ m$. This is much too small for you to observe any wave effects, such as diffraction or interference.

Waves or particles?

We cannot say that light is waves, or particles. Sometimes light *behaves like* waves, sometimes it *behaves like* particles. The same is true for electrons (and any other particle). On the microscopic scale of electrons and photons, we discover that matter and radiation behave in a way that is a strange mixture of the two. We just have to learn when the wave picture gives the better explanation, and when the particle picture is better.

? *Quick check questions*

Planck constant $h = 6.63 \times 10^{-34}$ J s

1 In a diffraction experiment, what would happen to the speed of the electrons if the accelerating voltage were decreased? How would the diffraction pattern change?

2 In Einstein's explanation of the photoelectric effect, do the electrons behave as particles or as waves?

3 Calculate the de Broglie wavelength for a car of mass 500 kg travelling at $20\ m\ s^{-1}$. Use your answer to explain why cars don't exhibit wave-like behaviour.

4 Light has momentum. Each photon in a beam of light carries an amount of momentum that can be calculated using the de Broglie equation. For light of wavelength 700 nm, what is the momentum of each photon?

▶ Remember: momentum = mass × velocity.

Module 1: end-of-module questions

Speed of light in free space $c = 3.00 \times 10^8$ m s^{-1}

Planck constant $h = 6.63 \times 10^{-34}$ J s

Electron charge $e = 1.6 \times 10^{-19}$ C

1 Two isotopes of neon are represented by the symbols $^{20}_{10}$Ne and $^{22}_{10}$Ne. In what respect are these isotopes **a** the same, **b** different? (3)

2 a The nuclide carbon-12 is represented by the symbol $^{12}_{6}$C.

  i How many protons are in a carbon-12 nucleus?

  ii How many electrons are in a neutral atom of carbon-12? (2)

  b One isotope of nitrogen is nitrogen-14. In what way does nitrogen-14 differ from carbon-14? (1)

3 In the Rutherford alpha-particle scattering experiment, alpha-particles were projected at a thin metal foil in a vacuum. The alpha-particles were scattered in various directions.

  a In which direction were the majority of the alpha-particles detected? (1)

  b The diagram shows the paths of two alpha-particles A and B approaching the nuclei of atoms in a metal foil.

  On a copy of the diagram, complete the paths to show how the alpha-particles might be deflected. (2)

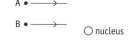

  c Explain why the metal foil should be thin. (1)

  d Explain why, in the apparatus used for the experiment, the alpha-particles and foil were in a vacuum. (1)

4 a Name a particle that may be classified as a *lepton*.

  b Hadrons may be classified as *mesons* or *baryons*. What is the essential difference between these two types of hadron?

  c The table shows values of charge, baryon number and strangeness for three quarks. A Σ^- (sigma-minus) quark consists of two down quarks and a strange quark (dds). What are its charge and its baryon number?

quark	charge Q	baryon number B	strangeness S
up, u	$+\frac{2}{3}$	$\frac{1}{3}$	0
down, d	$-\frac{1}{3}$	$\frac{1}{3}$	0
strange, s	$-\frac{1}{3}$	$\frac{1}{3}$	-1

d The following equation represents a suggested interaction between particles:

$$p + p \rightarrow p + p + n$$

Are charge and baryon number conserved in this reaction? Explain whether you would expect this reaction ever to be observed.

5 a What is the role of exchange particles in the interaction between elementary particles?

b What is the name of the exchange particle involved in the interaction represented by the Feynman diagram shown here?

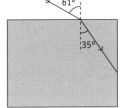

c Write an equation to represent this interaction.

6 The figure shows the path of a ray of light through a rectangular glass block.

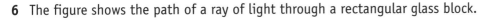

a Calculate the refractive index of the glass. (2)

b Calculate the critical angle for the glass–air boundary. (1)

c On a copy of the figure, sketch the path of the ray of light as it passes through the block and into the air. (2)

7 The refractive index of glass is different for different colours of light. Blue light is incident on a glass–air boundary. The refractive index of the glass is 1.53.

a Calculate:

i the speed of blue light in the glass (the speed of light in air is 3.00×10^{8} m s^{-1}),

ii the critical angle for blue light at the glass–air boundary. (3)

b The refractive index of the glass for red light is less than 1.53. State and explain:

i whether red light travels faster or slower than blue light in the glass,

ii whether the critical angle for red light is greater or less than the angle calculated in **a ii**. (4)

8 a A ray of light passes from glass of refractive index n_1 into liquid of lower refractive index n_2.

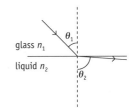

When the angle of incidence θ_1 is equal to the critical angle (θ_c), the angle of refraction θ_2 is 90°. Using the relationships $_1n_2 = n_2/n_1$ and $_1n_2 = \sin\theta_1/\sin\theta_2$, show that

$$\sin\theta_c = \frac{n_2}{n_1}$$ (3)

b The diagram shows a semicircular glass block that is separated from a horizontal surface by a liquid. The refractive index of the glass is 1.55.

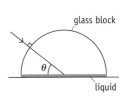

When a ray of light enters the glass block along a radius as shown, total internal reflection will occur at the glass–liquid boundary only if angle θ is less than 23°.

     i State the critical angle for the glass–liquid boundary. (1)

     ii Calculate the refractive index of the liquid. (2)

9 In an experiment to measure the work function of potassium, monochromatic ultraviolet radiation of wavelength 300 nm is shone on the metallic surface of some potassium. Electrons are emitted by the potassium. The fastest-moving electrons are found to have kinetic energy of 2.2 eV.

  a Calculate the energy of a single photon of the ultraviolet radiation, in eV. (3)

  b Explain what is meant by the term *work function*. (2)

  c Calculate the work function of potassium. (2)

  d Most of the electrons emitted by the potassium have a kinetic energy less than 2.2 eV. Explain why this is so. (1)

10 The graph shows how the maximum kinetic energy E_k of photo-electrons emitted from the surface of metal X varies with the frequency f of the radiation incident on the surface.

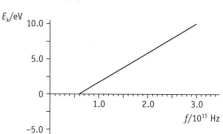

The equation relating E_k and f is

$$E_k = hf - \phi$$

  a Use the graph to determine

     i the work function ϕ of metal X, in eV, (2)

     ii the frequency of radiation that will just cause the emission of photoelectrons. (1)

  b On a copy of the graph draw a line to show the variation of E_k with f for photoelectrons emitted from the surface of a different metal having a greater work function than that of metal X. (2)

11 The diagram represents some of the electron energy levels for an atom. Level 0 is the ground state.

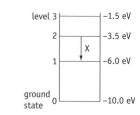

  a Express 3.5 eV in J. (1)

  b How many lines of the spectrum of the atom can be predicted from the energy levels shown? (1)

  c Calculate the frequency of the radiation emitted when an excited atom undergoes a transition from level 2 to level 1, labelled X on the diagram. (3)

  d The change in energy from level 2 to level 1 produces a green spectral line. Mark, with labelled arrows on the diagram, changes in energy level that could give rise to spectral lines corresponding to frequencies

     i less than that of the green spectral line (label this arrow A), (1)

     ii greater than that of the green line (label this arrow B). (1)

12 a Describe briefly the experimental evidence showing that electrons exhibit wave-like behaviour. (3)

  b Atoms may also show wave-like behaviour. Calculate the de Broglie wavelength of an atom whose momentum is 3×10^{-20} kg m s⁻¹. (3)

Module 2: Mechanics and molecular kinetic theory

This module contains ideas on forces and motion, and builds on your understanding of work, energy and power. Towards the end of the module we see the importance of considering gases as collections of molecules in random motion.

- **Block 2A** starts with the difference between vector and scalar quantities and goes on to show how we can analyse forces in structures.
- **Block 2B** shows how we describe the motion of an object in terms of displacement, velocity, acceleration and time.
- **Block 2C** considers forces and motion and introduces the important concept of momentum. Newton's laws of motion relate force to momentum and acceleration.
- **Block 2D** looks at work, energy and power, and the importance of the conservation of energy when applied to energy transfers.
- Finally, in **Block 2E**, we see how the gas equation (which relates the pressure, volume and temperature of a gas) can be explained by considering the motion of the gas molecules. This is called the kinetic theory, and this block looks at how the kinetic theory can be used to develop an understanding of what we mean by temperature.

Block 2A: Forces and equilibrium, pages 24–29

Ideas from GCSE	Content outline of Block 2A
<ul><li>Force diagrams</li><li>Moment of a force</li></ul>	<ul><li>Scalars and vectors</li><li>Adding and subtracting vectors</li><li>Components of a vector</li><li>Triangle of forces and resolving forces</li><li>Turning effect – moments and couples</li></ul>

Block 2B: Describing motion, pages 30–37

Ideas from GCSE	Content outline of Block 2B
<ul><li>Relationship between speed, distance and time</li><li>Graphical representation of speed, distance and time</li><li>Acceleration</li></ul>	<ul><li>Displacement, velocity and acceleration</li><li>Graphical representation of motion</li><li>Equations of motion</li></ul>

Block 2C: Explaining motion, pages 38–45

Ideas from GCSE	Content outline of Block 2C
<ul><li>Balanced forces do not alter velocity</li><li>Quantitative relation between force, mass and acceleration</li><li>Forces on a falling object</li></ul>	<ul><li>Momentum and its conservation</li><li>Collisions and explosions</li><li>Force, mass and acceleration</li><li>Motion under gravity</li></ul>

Block 2D: Work, energy and power, pages 46–51

Ideas from GCSE	Content outline of Block 2D
<ul><li>Work done, energy and power</li><li>Kinetic energy, gravitational potential energy</li><li>Energy transfers</li><li>Conservation of energy</li></ul>	<ul><li>Work, energy and power</li><li>Conservation of energy</li><li>Energy transfers</li><li>Specific heat capacity and specific latent heat</li></ul>

Block 2E: Molecular kinetic theory, pages 52–57

Ideas from GCSE	Content outline of Block 2E
<ul><li>Pressure as force/area</li><li>Boyle's law</li><li>Basic kinetic theory ideas</li></ul>	<ul><li>The Avogadro constant</li><li>The equation of state</li><li>Applying the kinetic theory</li><li>Temperature and kinetic energy</li></ul>

End-of-module questions, pages 58–60

Scalars and vectors

Scalar quantities have *magnitude* (*size*) only. Examples of scalar quantities are temperature, speed, distance and energy.

Vector quantities have both *magnitude* and *direction*. Examples of vector quantities are displacement, velocity, acceleration, force and momentum.

Adding two vectors

Vector quantities cannot simply be represented by a numerical value. Instead, to include their direction, they can be represented by drawing **vector diagrams**. In a vector diagram:
- the *length* of a line represents the magnitude of the vector quantity;
- the *direction* of a line represents the direction of the vector quantity.

Draw a **vector triangle** as follows.
- Choose a suitable scale. Draw a line (AB) to represent the first vector quantity. Add an arrow to the line.
- From the end of the first vector, draw a line (BC) to represent the second. Add an arrow.
- Return to the *start* of the first vector. Draw a line from this point to the *end* of the second vector. This line (AC) represents the **resultant** of the two. Add *two* arrowheads to show that this is the resultant of the other two vectors. This process is called **adding two vectors**.

If two vectors act in the same straight line, the resultant vector is found by adding the two, but you must take their direction into account.

> Vector diagrams are a type of scale drawing.

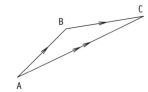

> There are two ways of drawing the triangle, but the resultant is exactly the same for both.

Worked example

Two tug boats pull a ship into a harbour. The forces in the towing cables are 85 kN and 120 kN as shown. Calculate the magnitude and direction of the resultant force.

The diagram shows how the two forces are added.
Step 1 Draw a vector to represent the 120 kN force.
Step 2 Draw a vector to represent the 85 kN force.
Step 3 Draw the resultant vector.
Step 4 Measure or calculate the resultant. In this case we have a right-angled triangle so we can calculate the resultant using Pythagoras' theorem.

$$R^2 = (120 \text{ kN})^2 + (85 \text{ kN})^2 = 14\,400 \text{ kN}^2 + 7225 \text{ kN}^2 = 21\,625 \text{ kN}^2$$

$$R = \sqrt{21\,625 \text{ kN}^2} = 147 \text{ kN}$$

To find the direction we need to find angle θ in the diagram.

$$\tan \theta = \frac{\text{opp}}{\text{adj}} = \frac{85}{120} = 0.708$$

$$\theta = \tan^{-1} 0.708 = 35.3°$$

The answers can be checked by drawing a scale diagram and measuring both the length of the resultant and the angle.

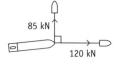

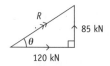

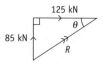

alternative triangle
same resultant

> ✓ *Quick check 1*

Resolving a vector

Sometimes it is useful to **resolve** (break down) a vector quantity into two **components** at 90° to one another.

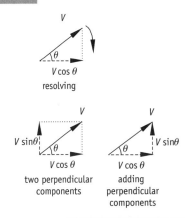
resolving

Imagine turning the vector *V* round to point in the direction of interest. If you turn the vector through an angle θ, the component of *V* in this direction is $V \cos \theta$.

A vector may be replaced by two perpendicular components whose values are $V \cos \theta$ and $V \sin \theta$. Notice that if these two components are *added* (see page 24), the resultant is the original vector.

two perpendicular components

adding perpendicular components

The *perpendicular components* of a vector are *independent* of one another. Changing one component has no effect on the other.

✓ *Quick check 2*

Worked example

A string is tied to a hook in a wall and is pulled with a force of 50 N at an angle of 30° to the wall. Calculate:

1 F_V, the vertical component of the force tending to bend the hook downwards;

2 F_H, the horizontal component of the force trying to pull the hook out of the wall.

Step 1 Sketch a diagram marking the forces and relevant angles.

Step 2 Calculate the vertical component:

$$F_V = 50 \text{ N} \times \cos 30° = 43.3 \text{ N}$$

Step 3 Calculate the horizontal component:

$$F_H = 50 \text{ N} \times \sin 30° = 25 \text{ N}$$

✓ *Quick check 2–4*

❓ Quick check questions

1 Find the single vector that is the resultant of the two forces shown in the upper diagram.

2 A loaded sledge (middle diagram) is pulled with a force of 200 N using a rope that is at an angle of 30° to the horizontal. Calculate the horizontal component of the pulling force.

3 Gliders need to be towed until they reach a suitable speed and height. A glider is pulled by a cable that has a tension force of 10 kN (lower diagram). If the cable makes an angle of 25° to the horizontal, calculate the horizontal and vertical components of the tension force.

4 A child is speeding at 8.0 m s⁻¹ down a water slide inclined at 35° to the horizontal. Calculate the horizontal and vertical components of the child's velocity.

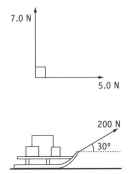

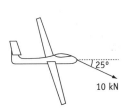

Equilibrium

When an object is in **equilibrium**, the forces on it *balance*. There is no tendency for it to move up or down, or to one side or the other. Also the object will not rotate. There is no tendency for the object to rotate if all the forces pass through a *single point* in the object.

▶▶ *More about rotation on pages 28–29.*

There are two ways of solving problems in which a point is in equilibrium. Either

- draw a **triangle of forces**, or
- **resolve** forces in two directions *at right angles*.

Triangle of forces

You can use this method if *three* forces act on an object that is in equilibrium, and if the forces all pass through the same point. You must know the directions of the forces and the magnitude of at least one. The method enables you to find the unknown force or forces. Use either a scale drawing or a sketch with calculations.

Step 1 Draw a diagram showing the three forces acting at a single point.

Step 2 Draw a line to represent the direction and size of a given force.

Step 3 Through one end of your line draw a second line parallel to one of the other forces.

Step 4 Through the other end of the first line draw a line parallel to the third force.

Step 5 Put arrows on the sides of your triangle to represent the directions of the forces. The arrows must follow one another nose-to-tail around the triangle.

Step 6 Find the unknown forces by measuring (if scale drawing) or calculation.

Worked example

A pendulum bob of weight 10 N is pulled to one side by a horizontal force *P*. Calculate the force *P* and the tension *T* in the string.

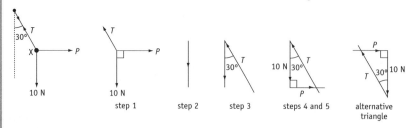

All three forces pass through X, which is in equilibrium.

The figure shows how steps **1** to **5** above are applied to this problem. The lengths of the sides of the triangle represent the sizes of the three forces. The arrows tell you the directions of the unknown forces.

Now apply step **6**, using trigonometry to find the unknown forces:

$$\tan 30° = \frac{P}{10\ N}$$

$$P = 10 \text{ N} \times \tan 30° = 5.8 \text{ N}$$
$$\cos 30° = \frac{10 \text{ N}}{T}$$
$$T = \frac{10 \text{ N}}{\cos 30°} = 11.5 \text{ N}$$

If you've made a scale drawing, you can simply measure the lengths of the sides and convert to newtons.

In step **3** it doesn't matter through which end of your first line you draw the second line. Check that if you had drawn force *T* through the bottom end of the 10 N force you would have ended up with the alternative triangle. The calculations based on this triangle would give the same answers for forces *T* and *P*.

✓ *Quick check 1, 3*

Solution by resolving forces

Resolve forces in two directions and use the rule that the *resultant* force in each of the two directions is zero. For example, upward forces = downward forces, and forces to left = forces to right.

Worked example

Consider the above problem using the triangle of forces. Calculate forces *P* and *T*.

Step 1 Resolve forces vertically and horizontally.
vertical component of *T* = *T* cos 30° upwards
horizontal component of *T* = *T* sin 30° to the left

Step 2 Resolving vertically:
$$T \cos 30° \text{ (upwards)} = 10 \text{ N (downwards)}$$
$$T = \frac{10 \text{ N}}{\cos 30°} = 11.5 \text{ N}$$

Step 3 Resolving horizontally:
$$T \sin 30° \text{ (to the left)} = P \text{ (to the right)}$$
Using the value for *T* from step **2**:
$$P = 11.5 \text{ N} \times \sin 30° = 5.8 \text{ N}$$

✓ *Quick check 2, 4*

? *Quick check questions*

1 Point X is in equilibrium under the action of the three forces shown. Draw a triangle of forces and determine the magnitudes of forces F_1 and F_2.

2 Confirm your answers to question **1** by using the technique of resolving forces to determine F_1 and F_2.

3 A tightrope walker of weight 650 N is at the midpoint of the stretched cable (see diagram at foot of page). The cable makes an angle of 8° to the horizontal each side. Draw a triangle of forces and determine the tension in the cable.

4 A child on a rope swing is pulled back by a horizontal force until the swing is at an angle of 45° to the vertical. The child weighs 400 N. Resolve forces vertically to calculate the tension in the rope and horizontally to find the pulling force.

▶ Don't confuse the triangle of forces with any triangles in the problem or structure. It helps to draw a small circle around the point that is in equilibrium and to deal only with the forces in the circle.

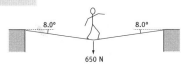

Turning effects

Moment of a force

Forces can have many different effects on the objects they act on. The **moment** of a force tells us about its *turning effect*.

The moment of a force about a point (the *pivot*) is defined as:

> **moment = *magnitude* of the force × perpendicular *distance* of its line of action from the pivot**

It is important to be able to determine the distance between the point and the line of action (direction) of the force. The worked example illustrates another method.

Units   Moment is measured in **newton-metres** (N m).

Note that 1 N m is not the same as 1 J. In calculating work done (in J), force and distance are in the *same* direction. In calculating moment (in N m), force and distance are *perpendicular*.

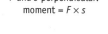

F and s perpendicular:
moment = $F \times s$

Draw a line from point P at 90° to line of force:
moment = $F \times s$

Worked example

A force of 100 N acts at an angle of 30° to a beam, and at a distance x = 4.0 m from one end. What is the moment of the force about this end?

Method 1
Step 1   Draw the line of action of the force. Then draw a perpendicular from P to the line of action.

Step 2   Calculate the length s of this line:

$$s = 4.0 \text{ m} \times \sin 30° = 2.0 \text{ m}$$

Step 3   Multiply by the force to find the moment:

$$\text{moment} = F \times s = 100 \text{ N} \times 2.0 \text{ m} = 200 \text{ N m}$$

Method 2
Step 1   Calculate the component of F perpendicular to x:

$$\text{component of } F = 100 \text{ N} \times \cos 60° = 50 \text{ N}$$

Step 2   Calculate the moment of this component about P:

$$\text{moment} = 50 \text{ N} \times 4.0 \text{ m} = 200 \text{ N m}$$

Of course, both methods give the same answer.

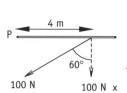

✓ *Quick check 1*

Torque of a couple

A **couple** is a pair of forces. They are equal in magnitude, and act in opposite directions, but they do not lie in the same line. Because of their equal and opposite sizes, they do not make the object accelerate away. However, because they do not line up, they tend to make the object *rotate*. The moment of a couple is known as its **torque** (units: N m).

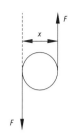

> **torque of a couple = magnitude of one force × perpendicular distance between them**
> **torque = $F \times x$**

✓ *Quick check 2*

Equilibrium

If an object is **in equilibrium**, it will not accelerate and it will not start to rotate. For this to happen:

- there must be no resultant (unbalanced) force acting on it;
- there must be no resultant torque acting on it. This means that:

> **sum of anticlockwise moments = sum of clockwise moments**

✓ *Quick check 3*

Centre of gravity and centre of mass

An object may have a complicated shape; gravity acts on all parts of it. Every object has a single point, called its **centre of gravity**, on each side of which the moments of all the separate parts of the object are balanced.

The **centre of mass** of an object is the single point at which all the *mass* can be considered to be concentrated. For small objects in a uniform gravitational field, the centre of mass and the centre of gravity coincide.

We can represent the weight of an object by a downward arrow acting at its centre of gravity or centre of mass.

centre of gravity

weight

| total anticlockwise moment of this part | = | total clockwise moment of this part |

? Quick check questions

1 Calculate the moment about P of each of the two forces shown in the upper diagram.

2 Which two of the forces shown in the lower diagram constitute a couple? What is their torque?

3 Copy the diagram for question **2**. Add another force, acting at point X, that will leave the object in equilibrium. Show that there is no resultant force or torque acting on the object.

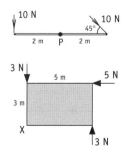

Velocity and displacement

When an object moves, we may be able to describe its motion using a graph, or an equation. First, we need to define some basic terms.

Speed, distance, time

We can find the **average speed** of a moving object by measuring the distance it travels in an interval of time:

$$\text{average speed} = \frac{\text{distance travelled}}{\text{time taken}}$$

This can only tell us its *average* speed; it may be speeding up or slowing down.

✓ Quick check 1

Motion in a straight line

If an object is moving at a steady speed in a straight line, it is in **uniform motion**. Two quantities describe its motion:

- **displacement** – the distance it has travelled in a particular direction;
- **velocity** – its speed in a particular direction.

These are related by the equation:

$$\text{velocity} = \frac{\text{displacement}}{\text{time}} \qquad v = \frac{s}{t}$$

It is important to be able to rearrange the equation for velocity to make time or displacement the subject:

$$\text{displacement} = \text{velocity} \times \text{time} \qquad s = vt$$

$$\text{time} = \frac{\text{displacement}}{\text{velocity}} \qquad t = \frac{s}{v}$$

> ❿ Take care! The symbol s is used for displacement, not speed. Do not confuse it with s for seconds.

SI units In the international system of units (SI units), displacement or distance is measured in **metres** (m), time in **seconds** (s) and velocity in **metres per second** (m s^{-1}). You may come across velocities in a variety of units; keep an eye on the units of displacement and time:

$$\text{m s}^{-1} \quad \text{mm s}^{-1} \quad \text{km s}^{-1} \quad \text{km h}^{-1} \quad \text{km y}^{-1}$$

> ❿ In the SI system, metres and seconds are fundamental or base units.

Worked example

A car travels at 25 m s^{-1} for 5 minutes due north along a straight road. What is its displacement after this time?

Step 1 Write down what you know, and what you want to know:

velocity v = 25 m s^{-1}, time t = 5 min = 300 s, displacement s = ?

Step 2 Choose the form of the equation with displacement as its subject:

$$\text{displacement} = \text{velocity} \times \text{time} \qquad s = vt$$

Step 3 Substitute values and solve:

$$s = 25 \text{ m s}^{-1} \times 300 \text{ s} = 7500 \text{ m} = 7.5 \text{ km}$$

So the car's displacement is 7.5 km due north. Note that, to give a complete answer, we have included the *direction* of the displacement.

✓ *Quick check 2, 3*

Vector and scalar quantities

The definitions of displacement and velocity should remind you of the difference between vector and scalar quantities.

- A **vector quantity** has both magnitude (size) and direction (e.g. displacement, velocity).
- A **scalar quantity** has just magnitude (e.g. distance, speed).

▶▶ *More about the representation of vector quantities on page 24.*

More about the representation of vector quantities on page 24.

✓ *Quick check 4*

Displacement–time graphs

The shape of an object's **displacement–time graph** shows how its motion is changing. In this example,

1 a straight line sloping up shows that it is going away at a steady speed ('*positive*' velocity);

2 a horizontal line shows that it is stationary for a while ('*zero*' velocity);

3 a straight line sloping down shows that it is coming back at a steady speed ('*negative*' velocity).

In the curved graph the object is speeding up (accelerating); its velocity is positive and increasing.

The **gradient** (slope) of the displacement–time graph is the velocity. Here, Δ (delta) just means 'change in', so Δ*s* means 'change in displacement'.

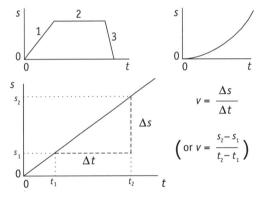

$$v = \frac{\Delta s}{\Delta t}$$

$$\left(\text{or } v = \frac{s_2 - s_1}{t_2 - t_1} \right)$$

▶ Check that the y-axis shows displacement (or distance).

✓ *Quick check 5*

? *Quick check questions*

1 A bus travels along its 20 km route in 40 minutes. Calculate its average speed, and explain why this is only an average.

2 A rocket rises 2000 m vertically upwards in 10 s. What is its average velocity?

3 How long will it take a car travelling at 30 m s^{-1} to travel 1200 m?

4 A spacecraft orbits the Earth at a constant speed of 8 km s^{-1}. Explain whether its velocity is also constant.

5 The table shows how the displacement of a runner along a straight track changes with time. Plot a displacement–time graph. Use it to find the runner's velocity during the first 10 s.

displacement/m	0	45	90	130	172
time/s	0	5	10	15	20

Acceleration

An object is **accelerating** if it is speeding up, **decelerating** if it is slowing down. Its acceleration or deceleration is a measure of how rapidly its velocity is changing.

Defining acceleration

The **acceleration** of an object is the rate of change of its velocity. If its velocity v changes by an amount Δv in a time interval Δt, its acceleration a is given by:

$$\text{acceleration} = \frac{\text{change in velocity}}{\text{time taken}} \qquad a = \frac{\Delta v}{\Delta t}$$

> Here, Δ (delta) does not represent a quantity. It stands for 'a change in'. So Δv means 'change in velocity'.

We can write this in a different way:

$$\text{acceleration} = \frac{\text{final velocity} - \text{initial velocity}}{\text{time}} \qquad a = \frac{v - u}{t}$$

> We need two symbols for velocity. Remember that u comes before v, so u represents initial velocity.

Units Acceleration is almost always given in m s^{-2} (metres per second squared). It can help to think of an acceleration of, say, 10 m s^{-2} as an increase in velocity of 10 m s^{-1} every second.

Signs An object with *negative* acceleration (deceleration) is slowing down. An object with *zero* acceleration either has uniform velocity or is stationary. *Positive* acceleration means speeding up.

Worked example

A car accelerates from 10 m s^{-1} to 18 m s^{-1} in 4 s. What is its acceleration?

Step 1 Write down what you know, and what you want to know:

$$u = 10 \text{ m s}^{-1}, \quad v = 18 \text{ m s}^{-1}, \quad t = 4 \text{ s}, \quad a = ?$$

Step 2 Write down the equation, substitute and solve:

$$a = \frac{v - u}{t} = \frac{18 \text{ m s}^{-1} - 10 \text{ m s}^{-1}}{4 \text{ s}} = \frac{8 \text{ m s}^{-1}}{4 \text{ s}} = 2 \text{ m s}^{-2}$$

> See the note on units in calculations on page iii.

> ✓ *Quick check 1, 2*

Typical values

It is useful to remember the following values.

● The acceleration in free fall is about 10 m s^{-2} (this is g – see pages 36 and 44).

● The acceleration of a car or a person is usually just a few m s^{-2}, as in the worked example.

Velocity–time graphs

Just as we can draw a displacement–time graph (see page 31), we can draw a **velocity–time graph** to show how an object's *velocity* is changing. In the first graph,

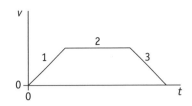

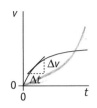

1 a straight line sloping up indicates steadily *increasing* speed (uniform '*positive*' acceleration);

2 a horizontal line shows steady speed ('*zero*' acceleration);

3 a straight line sloping down shows steadily *decreasing* speed ('*negative*' acceleration, or deceleration).

Curved graphs, like the one here, show changing or **non-uniform** acceleration (decreasing in this example).

$$a = \frac{\Delta v}{\Delta t}$$

$$\left(\text{or } a = \frac{v_2 - v_1}{t_2 - t_1} \right)$$

The **gradient** (slope) of the velocity–time graph is the acceleration. To find the acceleration at any instant, draw a *tangent* to the curve and calculate the gradient of the tangent.

The **area** under the velocity–time graph gives the displacement, because it is the average speed multiplied by the travelling time (see page 34).

> ▶ Always check the label on the y-axis – does the graph show displacement or velocity?

> ✓ *Quick check 2*

Worked example

Find the displacement up to 50 s of the object whose velocity–time graph is shown in the diagram.

Step 1 Divide the area under the graph into a triangle and a rectangle.

Step 2 Calculate the area of each part – see graph.

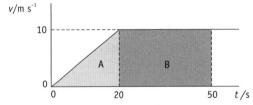

Triangle A: $\frac{1}{2}$ base × height = $\frac{1}{2}$ × 20 s × 10 m s^{-1} = 100 m

Rectangle B: 30 s × 10 m s^{-1} = 300 m

Step 3 Add together to give total displacement = 400 m.

> ✓ *Quick check 3*

❓ Quick check questions

1 An aircraft accelerates from 200 m s^{-1} to 300 m s^{-1} in 25 s. What is its acceleration?

2 A car has an acceleration of 8.0 m s^{-2}. How long will it take to reach a speed of 24 m s^{-1}, starting from rest?

3 A train travels at a steady speed of 30 m s^{-1} for 100 s. As it approaches a station, it decelerates at a steady rate so that it comes to a halt after 50 s. Draw a velocity–time graph for the train's motion, and use it to calculate

  a the train's acceleration as it slows down;

  b the total distance travelled during the time described.

> ▶ You will have to rearrange the equation for acceleration. Alternatively, think of the acceleration as '8.0 m s^{-1} every second'.

The equations of motion

The **equations of motion** can be used when an object is accelerating at a steady rate, i.e. its acceleration a is constant. There are *four* equations that need to be learnt. They link five quantities:

u initial velocity v final velocity s displacement

a acceleration t time

The four equations

1	$v = u + at$	This is a rearrangement of $a = \dfrac{v - u}{t}$ (see page 32)
2	$s = \dfrac{v + u}{2} \times t$	This says displacement = average velocity × time
3	$s = ut + \frac{1}{2}at^2$	With zero acceleration, this becomes displacement = velocity × time
4	$v^2 = u^2 + 2as$	This is easier to understand in terms of energy and work (see page 46)

These equations only apply to an object moving in a straight line with uniform (constant) acceleration.

✓ *Quick check 1*

Falling objects accelerate under the force of gravity. The acceleration due to gravity is called g.

> **For an object falling freely close to the Earth's surface,**
> **acceleration $a = g = 9.8$ m s^{-2}, approximately**

Worked examples

1 A car travelling at 20 m s^{-1} accelerates at 2 m s^{-2} for 5 s. How far will it travel in this time?

  Step 1   Write down what you know, and what you want to know:

$$u = 20 \text{ m s}^{-1}, \quad t = 5 \text{ s}, \quad a = 2 \text{ m s}^{-2}, \quad s = ?$$

  Step 2   Choose the appropriate equation linking these quantities:

$$s = ut + \tfrac{1}{2}at^2$$

  Step 3   Substitute and solve:

$$s = [20 \text{ m s}^{-1} \times 5 \text{ s}] + [\tfrac{1}{2} \times 2 \text{ m s}^{-2} \times (5 \text{ s})^2]$$

$$= 100 \text{ m} + 25 \text{ m} = 125 \text{ m}$$

Notice that the car's displacement is made up of two parts: 100 m is the distance it would have travelled in 5 s at a steady 20 m s^{-1}; 25 m is the extra distance travelled because it is acccelerating.

You may find it easier to omit the units, though they do provide a check that the quantities are correct. Always include a unit in the final answer.

2 For the car in worked example **1**, use equation 4 to find its velocity after it has travelled 125 m (after 5 s). Then use equation 1 to check your answer.

Step 1 Write down what you know, and what you want to know:

$u = 20$ m s^{-1}, $a = 2$ m s^{-2}, $s = 125$ m (from worked example 1), $v = ?$

Step 2 Choose the appropriate equation linking these quantities. The question requires equation 4:

$$v^2 = u^2 + 2as$$

Step 3 Substitute and solve:

$$v^2 = (20 \text{ m s}^{-1})^2 + [2 \times 2 \text{ m s}^{-2} \times 125 \text{ m}]$$
$$= 400 \text{ m}^2 \text{ s}^{-2} + 500 \text{ m}^2 \text{ s}^{-2} = 900 \text{ m}^2 \text{ s}^{-2}$$
$$v = \sqrt{900 \text{ m}^2\text{s}^{-2}} = 30 \text{ m s}^{-1}$$

Step 4 Check using equation 1 ($t = 5$ s):

$$v = u + at = 20 \text{ m s}^{-1} + (2 \text{ m s}^{-2} \times 5 \text{ s}) = 30 \text{ m s}^{-1}$$

> Note that it is necessary to start a new line when changing from v^2 to v.

> ✓ *Quick check 2–6*

Checking units

In calculations, both numbers and units contribute to the answer. By checking units as you go along, you will have a useful check that you are calculating correctly. For example, in Step 3 of worked example **1** above, we have two terms added together:

$$s = [20 \text{ m s}^{-1} \times 5 \text{ s}] + [\tfrac{1}{2} \times 2 \text{ m s}^{-2} \times (5 \text{ s})^2]$$

First term: units are m s^{-1} × s = m (because s^{-1} and s cancel).

Second term: units are m s^{-2} × s^2 = m (because s^{-2} and s^2 cancel).

Quick check questions

1 An aircraft accelerates at a steady rate from 200 m s^{-1} to 300 m s^{-1} in 80 s. Calculate its acceleration in this time, and its average speed.

2 A stone drops from rest with an acceleration of 9.8 m s^{-2}. How far will it fall in 2.0 s?

> 'From rest' tells you that its initial velocity was zero.

3 A skier moving at a steady speed of 15 m s^{-1} reaches a steeper slope where her acceleration is 1.25 m·s^{-2}. How fast will she be travelling after she has moved 160 m from the top of the steeper slope?

4 A train travelling at 10 m s^{-1} accelerates steadily. After 45 s it has reached a speed of 14 m s^{-1}. How far does it travel in this time?

5 At the start of a race, a runner accelerates from rest with a uniform acceleration of 4.5 m s^{-2} for 1.8 s. How fast will she be moving after this time?

6 At a motorway exit, a driver brakes from 30 m s^{-1} to 12 m s^{-1} with a deceleration of 2 m s^{-2}. For how long and over what distance is he braking?

> Here a is negative.

Independence of vertical and horizontal motion

When you throw a ball to another person, the ball moves upwards and forwards at the same time. The ball is a **projectile**. Sometimes we need to apply the equations of motion to objects moving forwards (horizontally) and rising or falling (vertically) at the same time.

▸▸ *You will need a knowledge of resolving vectors for this – see page 25.*

Vertical and horizontal components

If you drop a ball, it falls to the ground under the force of gravity at an acceleration, called g, of about 9.8 m s^{-2}. If you throw the ball horizontally, level with the ground, from the same height, it will fall with the *same* vertical acceleration of 9.8 m s^{-2}, so will take the same time to reach the ground. Its horizontal **component** of velocity remains constant even though it is accelerating downwards. (In practice, drag or air resistance slows objects down – see page 45 – but to simplify things here we neglect air resistance.)

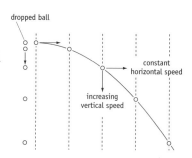

To summarise, for all projectiles:

> *vertical* **component of velocity increases with acceleration** $g = 9.8 \text{ m s}^{-2}$
> *horizontal* **component of velocity remains constant**

When dealing with projectiles, always remember to treat the vertical and horizontal motions *separately*.

> ▶ You don't need to remember complicated formulae by heart – the equations of motion are all that is required.

Worked example

An Arsenal defender kicks a football so that it leaves his boot at a velocity of 19.0 m s^{-1} at an angle of $30°$ to the ground. Determine whether it will reach an Arsenal midfielder 35 m away, without bouncing. Neglect air resistance.

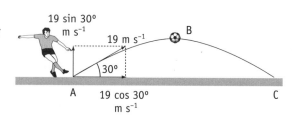

Step 1   Draw a sketch with arrows showing the vertical and horizontal components of the velocity (see 'Resolving a vector', page 25). Note that the motion is symmetrical about the highest point, B.

> vertical component = $19 \sin 30° \text{ m s}^{-1} = 9.5 \text{ m s}^{-1}$
> horizontal component = $19 \cos 30° \text{ m s}^{-1} = 16.5 \text{ m s}^{-1}$

✓ *Quick check 1*

Step 2 To find the horizontal distance travelled, we need the time taken from A to C. We will use the fact that the time t to reach B is half the time to reach C. To find t, consider vertical motion only and use $v = u + at$.

initial velocity = u = 9.5 m s^{-1}, final velocity = v = 0
acceleration = a = $-g$ = -9.8 m s^{-2}

The minus sign is needed because the ball is *decelerating* as it moves upwards. Applying $v = u + at$,

$$0 = 9.5 \text{ m s}^{-1} - 9.8 \text{ m s}^{-2} \times t$$

$$t \times 9.8 \text{ m s}^{-2} = 9.5 \text{ m s}^{-1}$$

$$t = 0.97 \text{ s}$$

The time for the ball to travel from A to C is therefore 2 × 0.97 s = 1.94 s.

Step 3 Consider horizontal motion only. Use *distance = horizontal speed × time* to find the distance AC (sometimes called the **range**):

distance = 16.5 m s^{-1} × 1.94 s = 32.0 m

The ball will not reach the midfielder directly, but it might roll once it hits the ground!

> ▶ At B the ball has *zero* vertical velocity because it has stopped moving upwards and is about to fall.

✓ *Quick check 2–4*

? *Quick check questions*

Acceleration due to gravity = 9.8 m s^{-2}.

1 A ball is thrown at a velocity of 18 m s^{-1} at an angle of 25° to the horizontal. Calculate the vertical and horizontal components of the velocity of the ball.

2 A darts player throws a dart horizontally towards the centre of a dartboard 2.0 m away. The speed of the dart as it leaves the player's hand is 12 m s^{-1}. Calculate the time taken for the dart to reach the board and the distance below the centre at which the dart hits the board.

3 In an ornamental fountain, a jet of water issues from a nozzle inclined at an angle of 80° to the horizontal. If the speed of the water as it leaves the nozzle is 10 m s^{-1}, determine the height above the level of the nozzle reached by the fountain.

4 For the fountain in question **3**, determine the range of the water jet, i.e. the maximum horizontal distance reached by the water at the level of the nozzle.

> ▶ Consider horizontal motion first.

Momentum

In physics the word *momentum* has a very particular meaning. The momentum of an object depends on its mass and its velocity.

Defining momentum

The **momentum** p of an object is defined as the product of its mass m and its velocity v:

$$p = m \times v$$

An object has momentum in a particular *direction*. Hence momentum is a *vector* quantity. For example, the momentum of a woman of mass 60 kg running at 8.0 m s^{-1} due north is

$$p = m \times v = 60 \text{ kg} \times 8.0 \text{ m s}^{-1} = 480 \text{ kg m s}^{-1} \text{ due north}$$

The units of p are simply the units of m and v multiplied together: kg m s^{-1}. There is no special name for this unit.

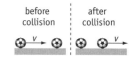
✓ *Quick check 1, 2*

Conservation of momentum

Like energy, momentum is a quantity that is *conserved*; that is, in any event, the total amount of momentum before the event is the same as the total amount after the event. This is made use of in solving problems – see pages 40–41. Here are some examples of situations to illustrate this idea. In each case, it is important to identify the **isolated system** for which momentum is conserved.

One ball rolls along and strikes a second, identical ball. The first stops dead, the second moves off with the speed of the first one. The momentum of the first ball has been transferred to the second. Isolated system: the two balls.

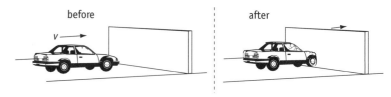
before collision | after collision

The car runs into a solid wall and stops dead. Where has its momentum gone? It has been transferred to the wall, and hence to the Earth, which moves *very* slightly faster to the right! Isolated system: car + Earth.

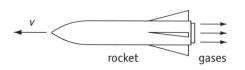
before | after

The rocket initially has zero momentum. When the rocket motor is fired hot gases are ejected at high velocity. For momentum to be conserved the momentum of the gases (backwards) must equal the momentum of the rocket (forwards). Isolated system: rocket + gases.

rocket | gases

✓ *Quick check 3*

Momentum and kinetic energy

Both momentum and kinetic energy depend on mass and velocity. This makes it very difficult to have separate mental images of these quantities. It is important to recall that:

- momentum is always conserved,
- the kinetic energy E_k of a mass m moving at speed v is $E_k = \frac{1}{2}mv^2$,
- kinetic energy is not always conserved.

Kinetic energy may be converted into other forms of energy, but momentum does not take different forms. A collision in which kinetic energy is conserved is described as **elastic**.

▶▶ *More on kinetic energy on page 48.*

✓ *Quick check 4*

? Quick check questions

1 Which of the following are *not* vector quantities: mass, velocity, momentum, kinetic energy?

2 A ship of mass 100 000 kg is sailing due west at 15 m s^{-1}. Calculate its momentum.

3 When a gun is fired, the bullet flies out very fast in one direction. The gun recoils more slowly in the opposite direction. Explain how momentum is conserved in this situation.

4 Which has more momentum, a boy of mass 40 kg running at 7.0 m s^{-1} or a girl of mass 32 kg running at 8.0 m s^{-1}? Which has more kinetic energy?

Collisions and explosions

Collisions and explosions are two examples of interactions between objects. Momentum is *conserved* in such interactions, and this is used to solve problems. If a collision is *elastic*, kinetic energy is also conserved, and this can help in calculations.

The types of collision used in the worked examples can be investigated in the laboratory using, for example, trolleys or a linear air track. You are expected to have had experience of analysing motion using data capture techniques with sensors such as light gates.

Collisions

We will only consider collisions in one dimension (i.e. along a line), but we could apply the same ideas to solve problems in two or three dimensions. Since momentum is conserved, we can write

▶ It is usually most helpful to start by drawing a pair of diagrams to show the situation before and after the interaction.

momentum before collision = momentum after collision

Worked examples

1 A trolley of mass 1 kg moving at 6 m s^{-1} collides with a second, stationary trolley of mass 2 kg. They stick together. With what velocity do they move off after the collision?

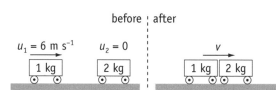

Step 1 Draw a before-and-after diagram; mark on it all the available information.

Step 2 Using momentum before collision = momentum after collision and substituting values gives

$$(1 \text{ kg} \times 6 \text{ m s}^{-1}) + (2 \text{ kg} \times 0 \text{ m s}^{-1}) = 3 \text{ kg} \times v$$

▶ You could omit 2 kg × 0 m s^{-1}, since this is obviously zero.

Step 3 Solve this equation for v:
$$6 \text{ kg m s}^{-1} = 3 \text{ kg} \times v$$
$$v = 2 \text{ m s}^{-1}$$

The velocity is in the direction the first trolley had been moving. It may be intuitively obvious that, since the mass increases by a factor of 3, the velocity decreases to a third of its initial value.

✓ *Quick check 1*

2 Two trolleys, each of mass 1 kg and moving at 4 m s^{-1}, collide head-on. They bounce apart; each has speed 2 m s^{-1} after the collision. Show that momentum is conserved. Is kinetic energy also conserved in this collision?

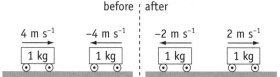

Step 1 Draw a before-and-after diagram; mark on it all the available information. Take care to give speeds to the right as positive, and to the left as negative.

Step 2 Here, we have to *show* that momentum is conserved. Calculate momentum before and after *separately*:

momentum before = (1 kg × 4 m s^{-1}) + (1 kg × −4 m s^{-1}) = 0 kg m s^{-1}
momentum after = (1 kg × −2 m s^{-1}) + (1 kg × 2 m s^{-1}) = 0 kg m s^{-1}

Since momentum before collision = momentum after collision, we have shown that momentum is conserved.

Step 3 Calculate the kinetic energy (using $\frac{1}{2}mv^2$) for each trolley, before and after the collision. Mark these values on the diagram. Note that all values are positive, since squaring gets rid of the minus signs.

kinetic energy before collision = 8 J + 8 J = 16 J
kinetic energy after collision = 2 J + 2 J = 4 J

So most of the kinetic energy disappears in the collision. (The trolleys may be deformed; heat and sound are produced.)

Explosions

Before an explosion, all parts of the system are at rest. Their combined momentum is zero. After the explosion, they are all flying apart. Each part has momentum, but their combined momentum, taking into account their different directions, is still zero.

Worked example

Two spring-loaded trolleys of masses 5 kg and 3 kg are stationary. When the spring is released, they fly apart. The lighter trolley moves at 4 m s^{-1}. How fast does the heavier one move?

Step 1 Draw a before-and-after diagram; mark on it all the available information.

Step 2 The total momentum after the explosion is zero, so we can write

momentum of trolley 1 + momentum of trolley 2 = 0

Step 3 Substitute values and solve for *v*:

$$(5 \text{ kg} \times v) + (3 \text{ kg} \times 4 \text{ m s}^{-1}) = 0$$
$$v \times 5 \text{ kg} = -12 \text{ kg m s}^{-1}$$
$$v = \frac{-12 \text{ kg m s}^{-1}}{5 \text{ kg}} = -2.4 \text{ m s}^{-1}$$

The minus sign means that trolley 1 is moving in the opposite direction to trolley 2. ✓ *Quick check 2,3*

? Quick check questions

1 A car of mass 500 kg travelling at 24 m s^{-1} collides with a second, stationary car of mass 700 kg. The two cars move off together. What is their shared velocity?

2 A marble of mass 20 g is moving to the right at 4 m s^{-1} when it collides with a smaller, stationary marble of mass 8 g. The smaller marble moves off to the right at 5 m s^{-1}. With what velocity (magnitude and direction) does the first marble move after the collision?

3 A cannon of mass 400 kg fires a shell of mass 20 kg. If the shell leaves the cannon at 300 m s^{-1}, with what velocity does the cannon recoil?

Newton's laws of motion

Newton's first law

- **An object continues in a state of rest or uniform motion in a straight line unless acted on by a resultant force.**

'Uniform motion' means constant velocity. A resultant force causes an object's velocity to change – it makes it accelerate.

✓ *Quick check 1*

Newton's second law

This law extends the first law to say what happens when an unbalanced or resultant force acts on an object. The force will change the object's momentum: the bigger the force, the greater the change in momentum per second.

- **When an unbalanced force acts on an object, its momentum changes; the rate of change of momentum is equal to the force producing it, and takes place in the direction of the force.**

This statement is a definition of what we mean by a **force** – something that causes a change in momentum. The bigger the force, the greater the change in momentum per second.

$$\text{force} = \text{rate of change of momentum} = \frac{\text{change of momentum}}{\text{time}}$$

$$F = \frac{\Delta(mv)}{\Delta t}$$

✓ *Quick check 2*

If the mass doesn't change, this can be written

$$F = m \times \frac{\Delta v}{\Delta t}$$

But $\Delta v / \Delta t = a = $ acceleration, so

> **force = mass × acceleration $F = ma$**

The object accelerates; the greater the force, the greater the acceleration. So we can also state Newton's second law as follows:

- **When an unbalanced force acts on an object, it accelerates; its acceleration is proportional to the unbalanced force, and takes place in the direction of the force.**

The equation $F = ma$ defines an object's **mass** as a measure of its *resistance to change in its motion*. Another word for mass in this sense is **inertia**.

✓ *Quick check 4, 5*

Worked example

A car of mass 1000 kg is acted on by two forces: a forward force of 500 N provided by its engine, and a retarding (backward) force of 200 N caused by air resistance. What is its acceleration?

Step 1   Draw a diagram to show the forces acting on the car. (It can help to draw a longer arrow for the larger force.)

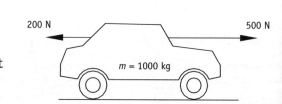

Step 2 Calculate the unbalanced force and note its direction:

$$F = 500 \text{ N} - 200 \text{ N} = 300 \text{ N forwards}$$

Step 3 Calculate the acceleration by rearranging $F = ma$:

$$a = \frac{F}{m} = \frac{300 \text{ N}}{1000 \text{ kg}} = \frac{300 \text{ kg m s}^{-2}}{1000 \text{ kg}} = 0.3 \text{ m s}^{-2}$$

So the car's acceleration is 0.3 m s^{-2} forwards.

> Don't forget to state its direction.

Units: the newton

The unit of force is the **newton**. The equation $F = ma$ defines the newton:

$$1 \text{ N} = 1 \text{ kg} \times 1 \text{ m s}^{-2}$$

A newton is the force that will give a mass of 1 kg an acceleration of 1 m s^{-2}. Equally, it will give a mass of 0.5 kg an acceleration of 2 m s^{-2}, and so on.

The kilogram, metre and second are **fundamental units** in the SI system. The newton is a **derived unit**. Most units we use are derived units; it is important to be able to trace them back to the fundamental units.

Newton's third law

Two objects that push or pull on one another exert equal and opposite forces on each other, sometimes referred to as *action* and *reaction*. The two forces must

- be equal in magnitude but opposite in direction,
- be of the same type (e.g. both contact forces, or gravitational, etc.),
- act on different objects.

The third law applies for any two objects interacting with one another – they do not *have* to be in contact.

- **When two objects interact, the forces they exert on each other are equal and opposite.**

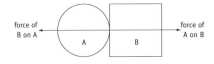

This is related to the conservation of momentum. Object A exerts force F on object B. This causes B's momentum to change. During the same time, B exerts force $-F$ on A, so A's momentum changes. The two changes in momentum are equal in magnitude but opposite in direction, because the forces are equal and opposite. If A's momentum increases by a certain amount, B's must decrease by an equal amount. Momentum is conserved.

> ✓ **Quick check 3**

? Quick check questions

1 If an object is in equilibrium, what can you say about the resultant force acting on it? What can you say about its velocity?

2 A rocket rises steadily upwards at a constant speed of 500 m s^{-1}. Its initial mass is 5×10^4 kg; after 10 minutes, this has decreased to 4.4×10^4 kg. Calculate the average force acting on the rocket during this time.

3 The diagram shows a person sitting still on a chair. Three forces are shown. Which pair must be equal in magnitude because of Newton's third law? Which pair must be equal because of Newton's second law?

4 A parachutist of mass 80 kg and weight 800 N is acted on by an upward drag force of 960 N. What is his acceleration (including its direction)?

5 When force F acts on object A, it is given an acceleration of 13 m s^{-2}. When the same force F acts on object B, its acceleration is 14 m s^{-2}. Which has the greater mass, A or B?

> In question 2, first find the change in momentum.

contact force of chair on person

weight of person

contact force of person on chair

Gravity and motion

The Earth's gravitational pull on us causes us to have **weight**. This is a force that acts on us all the time, so that we hardly notice its existence. Because our weight is proportional to our **mass**, we tend to get the two ideas confused.

Gravity and acceleration

If an object falls freely, it accelerates downwards. For objects near the Earth's surface, the acceleration caused by gravity has the *approximate* value $g = 9.8$ m s^{-2}. This decreases the further you go from the Earth's centre, so the value of g varies over the Earth's surface. It is greater nearer the poles where the Earth is slightly flattened, and less nearer the equator, as well as at higher altitudes.

In each succeeding second, the object falls further. It is accelerating downwards.

$t = 0$

$t = 1$ s

$t = 2$ s

$t = 3$ s

✓ *Quick check 1*

Gravity and weight

The fact that a falling object accelerates shows that there must be a resultant force acting on it – its **weight**. An object's weight depends on two factors:

- its mass m (the greater the mass, the greater its weight),
- the gravitational field strength g.

> **weight = mass × gravitational field strength** $W = mg$

The **gravitational field strength** g tells you how many newtons of force pull on each kilogram of mass. It has the approximate value $g = 9.8$ N kg^{-1} (newtons per kilogram) near the Earth's surface.

The symbol g is used for two things – the acceleration due to gravity, and the gravitational field strength. 1 N kg^{-1} is the same as 1 m s^{-2}. On the surface of the Moon, gravity is much weaker. The field strength is 1.6 N kg^{-1}, so falling objects have an acceleration of 1.6 m s^{-2}.

✓ *Quick check 2*

Weight and mass

Weight is a force, measured in newtons. It is represented on diagrams by an arrow. The weight of an object depends on where it is, because gravitational field strength varies from place to place.

Mass is a property of an object. It tells us how much matter it is made of. It is a measure of resistance to change in motion. It does not vary from place to place.

✓ *Quick check 3*

Moving through air

An object moving through air experiences air resistance, also known as **drag**. This is a resistive force, opposing motion, and is always in the opposite direction to velocity.

Air resistance increases as an object moves faster. Eventually, for a falling object, air resistance equals the object's weight, and the forces are balanced. The object cannot go any faster; it has reached **terminal velocity**.

This explains why cars, for example, have a top speed. The forward force of the engine is matched by the backward force of air resistance. To go faster, the car would need an engine that provided a greater force; alternatively, the car could be redesigned to reduce air resistance.

Forces on a falling object

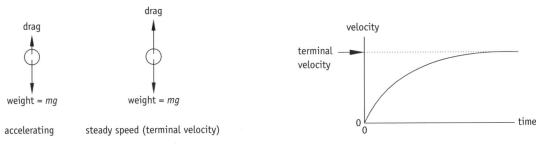

accelerating   steady speed (terminal velocity)

✓ **Quick check 4**

? Quick check questions

1 How far will an object fall, starting from rest, in 3 s? Its acceleration is 9.8 m s^{-2}.

2 The gravitational field strength on the surface of Mars is 3.8 N kg^{-1}. What will an object of mass 60 kg weigh there? How far will it fall, starting from rest, in 3 s?

3 Is weight a vector or a scalar quantity? Is mass a vector or a scalar quantity?

4 A skydiver is falling at a steady speed of 50 m s^{-1}. He opens his parachute so that the force of air resistance increases. Describe and explain how his motion will change after this.

Work, energy and power

We use forces to do things – to start things moving, to make them accelerate or decelerate, to change their shape. When a force changes the *energy* of something in this way, we say that it does **work**. Energy has been *transferred*.

> **work done = energy transferred**

Doing work

When a force F pushes an object for a distance s, the work W done by the force is

> **work done = force × displacement *in the direction of the force***
> $$W = Fs$$

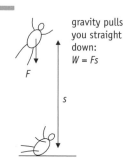

gravity pulls you straight down:
$W = Fs$

$W = Fs$ applies only when the displacement s is measured along the direction of the force. If s is at an angle θ to the direction of the force we use

$$W = Fs \cos \theta$$

Although both force and displacement in a particular direction are vector quantities, their product (work done) is a scalar.

Energy and work are both scalar quantities, measured in joules.

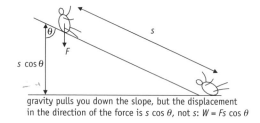

gravity pulls you down the slope, but the displacement in the direction of the force is $s \cos \theta$, not s: $W = Fs \cos \theta$

✓ *Quick check 1*

Defining the joule

Since work done tells us how much energy is transferred by a force, both work and energy are measured in the same units, called **joules** (J). The equation $W = Fs$ relates joules to newtons and metres.

> **1 joule = 1 newton × 1 metre   1 J = 1 N m**

1 joule is the energy transferred when a force of 1 newton moves through 1 metre. Equally, it is the energy transferred when a force of 0.5 N moves through 2 m, and so on. In fundamental units

$$1 \text{ J} = 1 \text{ N m} = 1 \text{ kg m s}^{-2} \text{ m} = 1 \text{ kg m}^2 \text{ s}^{-2}$$

✓ *Quick check 2*

Power

Power P is the *rate of doing work*, or the *rate of transferring energy*:

$$\text{power} = \frac{\text{work done}}{\text{time taken}} \qquad P = \frac{\Delta W}{\Delta t}$$

Power is measured in **watts**, W.

$$\text{1 watt = 1 joule per second} \qquad 1\ W = 1\ J\ s^{-1}$$

$$\text{1 kilowatt = 1 kW = 1000 J s}^{-1}$$

Since power = work/time, and work = force × distance,

$$\text{power} = \frac{\text{force} \times \text{distance}}{\text{time}} = \text{force} \times \text{velocity}$$

$$P = Fv$$

This is useful when working out the power needed to keep an object moving at a constant speed v against a resistive force F.

> Don't confuse ΔW (the symbol for work) with W (the unit for power).

> ✓ *Quick check 3*

Worked example

The engine of a boat develops a power of 28 kW when driving the boat at a constant speed of 6.0 m s^{-1}. Determine the total resistive forces opposing the motion of the boat.

Step 1 Write down what you know, and what you want to know:

$$P = 28 \times 10^3\ W, \quad v = 6.0\ m\ s^{-1}, \quad F = ?$$

Step 2 Rearrange the formula, substitute and solve:

$$F = \frac{P}{v} = \frac{28 \times 10^3\ W}{6.0\ m\ s^{-1}} = 4.7 \times 10^3\ N$$

since 1 W = 1 N m s^{-1}. So there is a total resistive force of 4.7 kN. Note that this is the same as the pushing force of the propeller because the boat is moving at constant speed (Newton's first law).

? Quick check questions

1 A car's engine provides a forward force of 500 N. It is opposed by a resistive force of 200 N. The car accelerates forwards for 20 m. How much work is done by each force? By how much does the car's energy increase?

2 Show that 1 J = 1 kg m^2 s^{-2}.

3 The engines of a light aircraft provide a power of 50 kW. How much energy do they transfer in 1 minute?

4 Which of the following are vector quantities: force, kinetic energy, gravitational potential energy, work done, power?

Energy transfers

Energy can be transferred from one place to another; for example, electricity can transfer energy from a power station to your home. Energy can also be transferred from one object to another; if you push a car to get it moving, you are transferring energy to the car.

Here we will consider *mechanical* **transfers of energy**, i.e. transfers involving forces. Energy can also be transferred in other ways, such as by heating and by electricity.

Conservation of energy

When energy is transferred from place to place, or converted from one form to another, the total amount always remains constant. This is the **principle of conservation of energy**. We make use of this in the worked example opposite.

However, during energy transfers and conversions, some of the energy may end up in a form that we did not want, such as heat or sound. The energy transfer or conversion is less than 100% efficient.

Kinetic energy

An object of mass m moving with velocity v has kinetic energy

$$E_k = \tfrac{1}{2}mv^2$$

Note that a change in velocity can contribute greatly to the change in kinetic energy – doubling the speed gives four times the energy.

✓ *Quick check 2*

Gravitational potential energy

When an object of weight mg is raised through a height Δh, work is done against gravity. The object's **gravitational potential energy** is increased by an amount ΔE_p given by

$$\text{gain in } E_p = \text{weight} \times \text{gain in height} \quad \Delta E_p = mg\,\Delta h$$

▶ The symbol Δ means 'a change in ...', so Δh means 'a change in height h'. It is important to think of Δh as a single mathematical symbol, not two quantities multiplied together.

E_p to E_k conversions

✓ *Quick check 2*

In many situations, an object's gravitational potential energy may be converted to kinetic energy, or vice versa. For example, when a ball rolls downhill, some of its gravitational potential energy is converted to kinetic energy.

Worked example

A stone falls from a height of 5 m. How fast is it moving when it reaches the ground?

Step 1   The decrease in the stone's gravitational potential energy E_p as it falls is equal to its gain in kinetic energy E_k:

$$mg \, \Delta h = \tfrac{1}{2}mv^2$$

Step 2   Cancel m from both sides:

$$g \, \Delta h = \tfrac{1}{2}v^2$$

Step 3   Substitute values and solve for v:

$$9.8 \text{ m s}^{-2} \times 5 \text{ m} = 0.5v^2$$

$$v^2 = 98 \text{ m}^2 \text{ s}^{-2}$$

$$v = 9.9 \text{ m s}^{-1}$$

The fact that m cancels out means that we would get the same answer for any value of m, i.e. all stones of whatever mass would fall at the same rate (neglecting air resistance).

✓ *Quick check 3, 4*

? Quick check questions

(Take $g = 9.8$ m s^{-2}.)

1   Are force and displacement scalar or vector quantities? Which of the following are vector quantities: kinetic energy, gravitational potential energy, work done, power?

2   A car of mass 1000 kg is travelling at 20 m s^{-1}. What is its kinetic energy? It climbs a hill 200 m high. By how much does its gravitational potential energy increase?

3   A person working on a television mast accidentally drops a spanner from a height of 150 m. Calculate the speed of the spanner when it reaches the ground. (Ignore air resistance.)

4   A girl on a swing is pulled back until she is 1.2 m above the lowest swing position, and then let go. Calculate her speed as she passes through the lowest position.

5   Consider the falling spanner in question **3**. In practice, it will be affected by air resistance. Use the ideas of work and conservation of energy to explain why the spanner will be moving more slowly than you have calculated in question **3**.

Specific heat capacity and specific latent heat

When a substance is heated, its temperature rises (unless it melts or boils). Some substances, such as metals, heat up more quickly than others – they have what is called a low *specific heat capacity*.

Energy must be supplied to melt or boil a substance. The temperature does not rise during such a *change of state*. The energy required can be calculated if the substance's *specific latent heat* is known.

Defining specific heat capacity

Energy must be supplied to raise the temperature of a substance. The amount of energy ΔQ that must be supplied depends on

- the mass of the substance, m,
- its specific heat capacity (s.h.c.), c,
- the temperature rise, $\Delta\theta$.

These four quantities are related by the equation

$$\Delta Q = mc\,\Delta\theta$$

Rearranging this equation gives

$$\text{specific heat capacity } c = \frac{\Delta Q}{m\,\Delta\theta} \qquad \text{s.h.c. = energy per kg per °C}$$

The **specific heat capacity** of a substance is the amount of energy that must be supplied to raise the temperature of 1 kg of the substance by 1 °C.

Units   Temperature rise $\Delta\theta$ is measured in **kelvin** (K). A rise of 1 K is the same as a rise of 1 °C. The units of specific heat capacity c are therefore J kg^{-1} K^{-1}.

▶ The word 'specific' means per unit mass or per kilogram.

Worked example

A 5 kg mass of water is heated electrically. A total of 210 kJ of energy is supplied. By how much will the temperature of the water rise? (Specific heat capacity of water = 4200 J kg^{-1} K^{-1}.)

Step 1   Write down what you know, and what you want to know:

$$\Delta Q = 210 \times 10^3 \text{ J}, \quad m = 5 \text{ kg}, \quad c = 4200 \text{ J kg}^{-1}\text{ K}^{-1}, \quad \Delta\theta = ?$$

Step 2   Rearrange the formula, substitute and solve:

$$\Delta\theta = \frac{\Delta Q}{mc} = \frac{210 \times 10^3 \text{ J}}{5 \text{ kg} \times 4200 \text{ J kg}^{-1}\text{ K}^{-1}} = 10 \text{ K}$$

So the water's temperature rises by 10 K (or 10 °C).

✓ *Quick check 1, 2*

Continuous flow

Sometimes we need to solve problems concerned with continuous flow heating or cooling. Examples are showers, hair driers and car radiators.

As the fluid (gas or liquid) passes over the heating element, energy is transferred to the fluid. From $\Delta Q = mc\,\Delta\theta$ it follows that

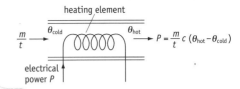

$$\textbf{power} = \textbf{energy per second} = \frac{\textbf{mass}}{\textbf{second}} \times \textbf{s.h.c.} \times \textbf{temperature change}$$

✓ **Quick check 3**

Heat exchangers

If a cooler substance, such as a fluid (liquid or gas), is used to cool a hotter one, the cooler one gains thermal energy while the hotter one loses thermal energy. If no energy is lost to, or gained from, the surroundings:

energy per second lost by hotter fluid = energy per second gained by cooler fluid

▶ For example, the water in a car radiator loses energy to the air flowing through the radiator. The water cools while the air is heated.

Defining specific latent heat

Melting, boiling, condensing and freezing are called *changes of state*. The amount of energy ΔQ that must be supplied to melt or boil a substance depends on
- the mass of the substance, m,
- its *specific latent heat* (s.l.h.), L.

These three quantities are related by the equation

$$\Delta Q = mL$$

Rearranging this equation gives

> **specific latent heat $L = \dfrac{\Delta Q}{m}$**   **s.l.h. = energy per kg**

▶ This equation does not involve a temperature rise $\Delta\theta$, since there is no change in temperature during a change of state.

The **specific latent heat** of a substance is the amount of energy that must be supplied to melt or boil 1 kg of the substance. L is measured in J kg^{-1}.

For a given substance, the s.l.h. for boiling or condensing (vaporisation) is greater than the s.l.h. for melting or freezing (fusion).

✓ **Quick check 4, 5**

? ## Quick check questions

1 How much energy must be supplied to raise the temperature of a 0.5 kg block of aluminium from 20 °C to 60 °C? (Specific heat capacity of aluminium = 900 J kg^{-1} K^{-1}.)

2 A 2.5 kg block of nylon at 20 °C is heated for 5 minutes using a 100 W heater. To what value will its temperature rise? (Specific heat capacity of nylon = 470 J kg^{-1} K^{-1}.)

3 A diesel engine is cooled by pumping water at 0.85 kg s^{-1} around the cylinders and through a heat exchanger (the 'radiator'). The water temperatures at the inlet and outlet of the radiator are 90 °C and 84 °C. Calculate the rate at which the cooling water loses energy in the radiator. (Specific heat capacity of water = 4.2 kJ kg^{-1} K^{-1}.)

4 How much energy must be supplied to vaporise 150 g of ethanol? (Specific latent heat of vaporisation of ethanol = 850 kJ kg^{-1}.)

5 A 2 kW electric kettle initially contained 800 g of water at 100 °C. It boiled dry in 15.0 minutes. Estimate the specific latent heat of vaporisation of water.

The equation of state

The **equation of state for an ideal gas** (also called the **ideal gas equation**) relates the pressure, volume and temperature of a gas. We can also relate these quantities to the average kinetic energy of the molecules of the gas. In these relationships, the amount of a gas is expressed in *moles* (see below).

Temperature scales

The **Celsius scale** of temperature was devised to give 100° between the melting point of water (0 °C) and its boiling point (100 °C). Temperatures in °C below the freezing point of water are therefore *negative*.

The **thermodynamic** (or **Kelvin**) **scale** is defined differently. One point is fixed: **absolute zero**, whose temperature is 0 K. The size of 1 degree on this scale is also fixed – it is an SI *fundamental* or *base* unit, and is the same as a degree Celsius. The thermodynamic scale does not depend on the physical properties of any particular substance.

a temperature rise of 1 K = a temperature rise of 1 °C

▶▶ *For the meaning of absolute zero, see page 56.*

Absolute zero is approximately –273 °C. To convert a temperature from one scale to the other:

°C = K − 273
K = °C + 273

> Remember that it is impossible to have negative temperatures on the Kelvin scale.

Symbols θ for temperatures in °C, T for temperatures in K.

✓ Quick check 1

The mole and the Avogadro constant

The **mole** (SI abbreviation: mol) is the unit of *amount* of a substance. One mole of any substance consists of a standard number of particles. This number is called the **Avogadro constant**, symbol N_A. It is the number of atoms in exactly 12 g of carbon-12.

$$N_A = 6.02 \times 10^{23} \text{ mol}^{-1}$$

The mass of 1 mole of any substance is the **relative molecular mass** of the substance, expressed in grams. For example:

- 1 mole of carbon-12 has a mass of 12 g and consists of 6.02×10^{23} molecules, i.e. its relative molecular mass is 12.

- The relative molecular mass of water, H_2O, is 18, so 1 mole of water has a mass of 18 g and also consists of 6.02×10^{23} molecules.

✓ Quick check 2

The equation of state for an ideal gas

The pressure p (in pascals, Pa), volume V (m^3) and absolute temperature T (K) of a gas are related by

$$pV = nRT$$

where n is the number of moles of the gas, and R is the **molar gas constant**, given by

$$R = 8.3 \text{ J K}^{-1} \text{ mol}^{-1}$$

An **ideal gas** is one that obeys this equation. In practice, most gases behave ideally only at low pressures and at temperatures well above their boiling points.

> ▶ Always remember to convert °C to K when using $pV = nRT$. Make sure V is in m^3, p is in Pa and n is in mol, not kg.

Worked example

What volume is occupied by 1 mole of a gas at a pressure of 1.0×10^5 Pa and a temperature of 273 K?

Step 1   Write down what you know, and what you want to know:

$$n = 1 \text{ mol}, \quad p = 1.0 \times 10^5 \text{ Pa}, \quad T = 273 \text{ K}, \quad R = 8.3 \text{ J K}^{-1} \text{ mol}^{-1}, \quad V = ?$$

Step 2   Rearrange the ideal gas equation, substitute and solve:

$$V = \frac{nRT}{p} = \frac{1 \text{ mol} \times 8.3 \text{ J K}^{-1} \text{ mol}^{-1} \times 273 \text{ K}}{1.0 \times 10^5 \text{ Pa}}$$

$$= 0.023 \text{ m}^3$$

When checking the units here, remember that 1 J = 1 N m (see page 46), and 1 Pa = 1 N m^{-2} (pressure = force per unit area). The mols and the Ks cancel, leaving J Pa^{-1}, so that, using SI units throughout, the volume is in m^3.

> ▶ If you're studying chemistry, you should recognise this as the volume of 1 mole at standard temperature and pressure, s.t.p.

> ✓ Quick check 3, 4

? **Quick check questions**

1   Convert:

   a   88 °C to K,

   b   253 K to °C.

2   How many particles are there in 5 mol of water? In 5 mol of uranium?

3   At what temperature will 10 mol of an ideal gas occupy 0.10 m^3 at a pressure of 2×10^5 Pa? (Molar gas constant $R = 8.3$ J K^{-1} mol^{-1}.)

4   An ideal gas initially occupies a volume of 10 litres. It is compressed at constant temperature so that its pressure increases by a factor of 2.5. Calculate its new volume.

Kinetic theory

The **kinetic theory of gases** was developed to relate the bulk properties of the whole gas (P, V and T) to the properties of its molecules. It helps to explain how individual fast-moving particles exert a steady pressure on the walls of the container of the gas.

We make certain assumptions about the way the molecules (particles) of the gas move and collide:

- The molecules are in rapid random motion, i.e. move haphazardly. One piece of evidence for this is **Brownian motion**.
- Their volume is negligible compared with the volume of the container of the gas.
- The molecules undergo perfectly elastic collisions (see page 39).
- They exert no forces on other molecules or the container walls except when they collide.
- The time spent in collision is negligible compared to the time between collisions.

A gas for which these assumptions hold true is called an **ideal gas**. Real gases behave almost as ideal gases at low pressures and at temperatures well above their boiling points.

The pressure is due to the particles colliding with the container walls. There are so many molecules that the walls experience a constant force due to their bombardment, rather than a series of occasional 'thumps'.

Model of an ideal gas

We will now look at the behaviour of an ideal gas of volume V m^3 containing N molecules. Consider the molecules in an *imaginary* tube at right angles to the container wall.

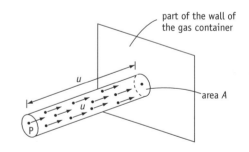

part of the wall of the gas container

area A

Suppose all the molecules in the tube move towards the wall with velocity u m s^{-1}. In 1 second all the molecules in a tube of length u metres will hit the wall. (A molecule P at the far end of the tube will just reach the wall in 1 s, so all those in front of it will have hit the wall.) On hitting the wall they bounce back.

momentum of a molecule before hitting wall = *mu*
momentum of the molecule after hitting wall = −*mu*

change of momentum = *mu* − (−*mu*) = 2*mu*

> ◗ Remember that momentum is a vector.

Suppose the number of molecules per m^3 is n, and the area at the end of the tube is A m^2, then:

volume of tube of length *u* metres = *A* × *u* m^3

number of molecules hitting wall each second = *n* × *A* × *u*

total change in momentum per second = 2*mu* × *nAu* = 2*mnAu*^{2}

But the change in momentum per second is the *force* of the molecules. This is Newton's second law (see page 42). Therefore:

$$\text{force} = 2mnAu^2$$

and this force acts over area A, so

$$\text{pressure } p = \frac{F}{A} = \frac{2mnAu^2}{A} = 2mnu^2$$

However, in a real gas:

- not all the molecules will be moving towards area A,
- the molecules have different speeds.

Whatever the direction of a single molecule, it will have components in three mutually perpendicular directions, each of which could be positive or negative (basically up/down, forward/back, left/right). So to find the pressure on A we need to divide our result by 6.

We must also replace u^2 by the *average* of all the values of u^2. This average value is written $\overline{c^2}$ and is called the **mean square speed**. The square root of $\overline{c^2}$ is called the **root mean square (r.m.s.) speed**.

Taking these two modifications into account, we get

$$p = \tfrac{1}{6} \times 2nm\overline{c^2} = \tfrac{1}{3}nm\overline{c^2}$$

Now $n = \dfrac{\text{number of molecules of gas}}{\text{total volume of gas}} = \dfrac{N}{V}$, so

$$pV = \tfrac{1}{3} Nm\overline{c^2}$$

Furthermore, Nm = (total number of molecules) × (mass of each molecule), which is the total mass of the gas. The density, ρ = mass/volume, so we can write

$$p = \tfrac{1}{3}\rho\,\overline{c^2}$$

✓ *Quick check 2, 3*

Worked example

Four particles have speeds 2 m s^{-1}, 4 m s^{-1}, 6 m s^{-1} and 8 m s^{-1}. Calculate their mean square speed.

▶ Note that the answer is not the same as the (mean speed)^2. Try it.

$$\overline{c^2} = \frac{2^2 + 4^2 + 6^2 + 8^2}{4} = \frac{120}{4} = 30 \text{ m}^2\text{ s}^{-2}$$

The root mean square speed is $\sqrt{30 \text{ m}^2\text{s}^{-2}} = 5.5$ m s^{-1}. This is different from the mean or average speed (5.0 m s^{-1}).

✓ *Quick check 1*

? Quick check questions

1 The speeds of five cars are 10 m s^{-1}, 12 m s^{-1}, 13 m s^{-1}, 20 m s^{-1} and 20 m s^{-1}. Calculate their mean speed and their root mean square speed.

2 The density of oxygen at a pressure of 101 kPa and a temperature of 273 K is 1.43 kg m^{-3}. Calculate the root mean square speed of oxygen molecules at this temperature and pressure.

3 The volume of a sample of ideal gas is reduced to one half and the pressure is increased to one and a half times its original value. Explain whether the r.m.s. speed of the gas molecules increases or decreases.

Internal energy

The **internal energy** of an object is simply the sum of the energies (kinetic and potential) of the molecules from which the object is made. The energy of each molecule keeps changing in a random fashion, due to collisions with its neighbours, but if we could take a snapshot of the system at an instant in time and add up the energies of all the molecules, we would find the internal energy. More formally:

The internal energy of a system is the sum of a random distribution of kinetic and potential energies associated with the molecules of the system

Energy can be transferred to a system in a variety of ways: by heating, by doing work, or electrically. The energy transferred is shared among the molecules of the system, increasing their energies. This increases the internal energy of the system.

✓ *Quick check 1*

Absolute zero

Cool a substance down, and its molecules move more and more slowly. At a temperature called **absolute zero** (0 K, 0 kelvin, equal to about −273 °C), the molecules have reached their lowest possible energy, though they don't quite stop moving even then. This is the temperature at which all substances have a minimum internal energy. It is impossible to cool anything below absolute zero.

Changes of state

Energy must be supplied to melt a solid, or to boil a liquid. As it *changes state*, the substance remains at a steady temperature. The energy input is increasing the potential energy of the molecules, not their kinetic energy. Bonds are broken between molecules, but they do not move any faster.

Temperature and molecular energy

The molecules of a gas have a spread or range of speeds – some are moving faster than others.

At high temperatures,

- the average speed of the molecules increases and there is a greater range of speeds,
- there are fewer molecules with any particular speed.

At any temperature, there are relatively few molecules with very low speeds, and relatively few with very high speeds.

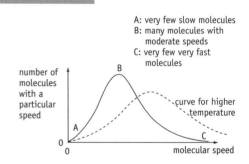

A: very few slow molecules
B: many molecules with moderate speeds
C: very few very fast molecules

The Boltzmann constant

In the derivation of $p = \frac{1}{3}\rho \overline{c^2}$ (page 55) we found that

$$pV = \tfrac{1}{3}Nm\overline{c^2}$$

This equation relates the bulk properties of the whole gas (p and V on the left-hand side) to the properties of its molecules (on the right-hand side). Applying this equation to the *equation of state* for an ideal gas (page 53) gives

$$\tfrac{1}{3}Nm\overline{c^2} = nRT$$

For 1 mol, $n = 1$ and $N = N_A$, the Avogadro constant (see page 52). So

$$\tfrac{1}{3}N_A m\overline{c^2} = RT$$

But $\tfrac{1}{3} = \tfrac{2}{3} \times \tfrac{1}{2}$, so we can write

$$\tfrac{2}{3} \times (\tfrac{1}{2}m\overline{c^2}) \times N_A = RT$$

$$\tfrac{1}{2}m\overline{c^2} = \frac{\tfrac{3}{2}RT}{N_A} = \tfrac{3}{2}(R/N_A) \times T$$

The left-hand side is the average kinetic energy of a molecule, which is therefore proportional to the absolute temperature:

$$E_k \propto T$$

R/N_A is called the **Boltzmann constant**, denoted by k, so

$$E_k = \tfrac{1}{2}m\overline{c^2} = \tfrac{3}{2}kT$$

The value of k is 1.38×10^{-23} J K^{-1}.

A *universal constant* is one that is the same everywhere in the universe. R and N_A are both universal constants, so $k = R/N_A$ is also a universal constant.

$$\text{for one } molecule,\ E_k = \tfrac{3}{2}kT$$
$$\text{for one } mol,\ E_k = \tfrac{3}{2}RT$$

✓ *Quick check 2–5*

Thermal equilibrium

Energy flows from a hot object to a cold object in contact with it. The temperature of the hot object falls and that of the cold object rises. The hot object loses (heat) energy and the cold one gains energy. Eventually there will be no net flow of energy from one to the other. This means that any energy flow from object 1 to object 2 will be exactly balanced by energy flow from object 2 to object 1. The objects are then said to be in **thermal equilibrium**. They will be at the *same temperature*.

? *Quick check questions*

Boltzmann constant $k = 1.38 \times 10^{-23}$ J K^{-1}
Avogadro number $N_A = 6.02 \times 10^{23}$
Universal gas constant $R = 8.3$ J K^{-1} mol^{-1}

1 A block of ice is cut exactly in half. What can you say about the internal energy of each half, compared to that of the original block?

2 The r.m.s. speed of atoms in a container of argon gas is 400 m s^{-1}. What would be the r.m.s. speed if the volume is doubled, with the temperature remaining the same?

3 What is the mean translational kinetic energy of a molecule of helium at a temperature of 37 °C?

4 The mean kinetic energy of a molecule of gas in a container is 8.75×10^{-21} J. What is the temperature of the gas?

5 The molar mass of helium is 4 g. Calculate, for helium gas at 100 K:
   a the internal energy of 1 mol of helium,
   b the internal energy of 1 g of helium,
   c the mean kinetic energy of one atom.

Module 2: end-of-module questions

Acceleration due to gravity g = 9.8 m s^{-2}

Molar gas constant R = 8.31 J K^{-1} mol^{-1}

Avogadro constant N_A = 6.02 × 10^{23} mol^{-1}

Boltzmann constant k = 1.38 × 10^{-23} J K^{-1}

1 The diagram shows a heavy load being dragged by two tractors. The tension in each cable is 20 kN.

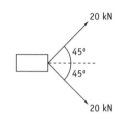

    a Draw a vector triangle. Use it to find the resultant of these two forces. (4)

    b The load moves at a steady speed along the ground. What can you say about the resistive force acting on it? (1)

2 The cable of an electrical power transmission system is supported at 200 m intervals by insulators attached to steel towers or pylons. The insulators hang vertically and the effective weight of the 200 m of cable supported by each insulator is 2.90 kN. The weight of the insulators is negligible compared to the weight of the cable.

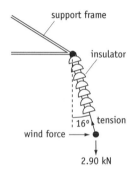

In a strong cross-wind, the insulators are deflected as shown.

For an angle of deflection of 16°, and by using a vector diagram or otherwise, determine

    a the tension in the insulator, (3)

    b the horizontal wind force on each metre of cable. (3)

3 a With the aid of a diagram, explain what is meant by the moment of a force. (2)

The diagram shows a diving board 4.00 m long that projects horizontally from point X. Its weight (2000 N) acts half way along its length, as shown. It is supported by a cable attached 1.00 m from the far end, that makes an angle of 50° with the vertical.

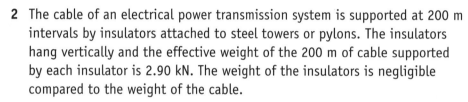

    b Calculate the moment of the board's weight about point X. (1)

    c Show that the tension T in the cable must be 2074 N if its moment is to counteract the moment of the board's weight. (2)

4 a What is the essential difference between speed and velocity? (1)

    b Which of the following quantities are vectors, and which are scalars? force, velocity, distance, acceleration, kinetic energy, power. (4)

5 a Write down a word equation that defines acceleration. (1)

    b The graph shows how the velocity of a car varied during part of a journey. Calculate the car's acceleration
      i during the first 10 s,

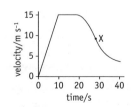

ii between 10 s and 20 s after the start,

iii at the instant X. (6)

6 In this question neglect air resistance and take the acceleration of free fall, *g*, as 10 m s^{-2}.

A projectile is fired with speed *v* at an angle of 39° to the ground as shown in figure 1.

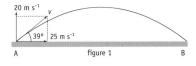
figure 1

The horizontal and vertical components of the velocity of the projectile are 25 m s^{-1} and 20 m s^{-1} respectively.

a Show that the time for the projectile to travel from A to B is 4.0 s. (2)

b Copy figures 2 and 3. On the axes sketch graphs to show the variation with time of the horizontal and vertical components of the projectile's velocity. Take upward motion as positive. (5)

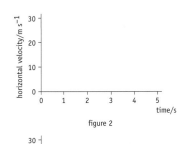
figure 2

c For the projectile, calculate
i the distance AB (i.e. the range),
ii the maximum height reached. (4)

d Calculate the initial speed *v* of the projectile. (2)

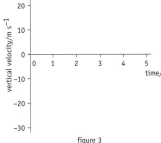
figure 3

7 Police investigating a road accident discover that one of the cars has bald tyres; most of the tread has been worn away. They suggest to the driver that he should have noticed problems when he attempted to make the car accelerate.

a Draw a diagram to show the horizontal forces that act between the road surface and the tyre of a car. Use your diagram to explain the origin of the motive force that causes a car to accelerate. (3)

b Suggest why a bald tyre will result in a reduced motive force on wet roads. (1)

c The driver claims that he was travelling at 20 m s^{-1} when he braked. The police say that he should have been able to stop in a distance of 40 m if his tyres had been roadworthy. What would the driver's deceleration have been if he had stopped in this distance? (2)

d Police tests show that the driver's bald tyres could not have provided a deceleration greater than 2.0 m s^{-2}. Determine the braking distance required if the car was travelling at 20 m s^{-1}. (2)

8 When the space shuttle comes in to land, it deploys a parachute to slow it down. The graph shows how the horizontal force *F* on the shuttle varies with time *t*.

a The parachute is deployed at point A on the graph. Explain why the force on the shuttle decreases in the region AB. (2)

b Copy the graph. On the same axes, add a further line to show how the shuttle's velocity changes. (2)

9 a Define (linear) momentum and state whether it is a vector or a scalar quantity. (2)

b Calculate the momentum of a bus of mass 6000 kg moving at 20 m s^{-1}. (1)

c State the principle of conservation of momentum. (2)

d Explain how the principle of conservation of momentum applies in the following situations:
   i a child, initially stationary, jumps up in the air;
   ii the child lands on the ground (without bouncing). (3)

10 a What is meant by an *elastic* collision? (1)

b A marble of mass 5.0 g moving at 1.0 m s^{-1} collides with an identical, stationary marble. The first marble stops dead and the second moves off at 1.0 m s^{-1}. Show that this collision is elastic. (3)

11 An electric wheelchair is to have a top speed of 4.0 m s^{-1} on level ground when the friction and air resistance forces are 18 N in total.

a Show that this specification would be met if the wheelchair is powered by an electric motor of output power 72 W. (2)

b A ramp outside a theatre is 25 m long and rises 1.0 m. For the wheelchair and driver of total weight 950 N, calculate:
   i the work done in raising the wheelchair and driver 1.0 m;
   ii the work done against friction and air resistance in travelling 25 m, given that the resistive force is unchanged at 18 N;
   iii the total work done;
   iv the time taken to climb the ramp if the output power of the motor is 72 W;
   v the speed of the wheelchair along the ramp. (7)

12 A fruit of mass 100 g falls from a branch of a tree, 4.00 m above the ground. It hits the ground with a speed of 8.86 m s^{-1}. Calculate:

a the change in the fruit's gravitational potential energy during the fall, (2)

b the change in its kinetic energy. (1)

c With reference to the principle of conservation of energy, state the energy transfers that occur as the fruit falls. (2)

13 a One mole of oxygen consists of 6.02×10^{23} molecules. Of what quantity is the mole the unit? (1)

b Write down an equation linking the pressure p and volume V of 1 mole of an ideal gas to its absolute temperature T. (1)

c Use your equation to calculate the pressure of 1 mole of oxygen at 100 °C if it occupies a volume of 0.025 m^3. (3)

14 a The equation $pV = \frac{1}{3}Nm\overline{c^2}$ relates the pressure p and volume V of a gas to its molecular properties. Explain the meanings of each of the symbols N, m and c^2. (3)

b A sample of 1.00 mole of a gas occupies a volume of 0.05 m^3 when its pressure is 5.00×10^4 Pa. Calculate
   i its temperature,
   ii the average kinetic energy of one of its molecules. (5)

Module 3: Current electricity and elastic properties of solids

There are three blocks in this module.

- **Block 3A** is the longest. It concerns electricity and will help you to solve problems involving current, voltage and resistance, as well as energy and power in electric circuits.
- **Block 3B** is short, and introduces some basic ideas about alternating currents and voltages. It shows how an oscilloscope is used to measure direct and alternating voltages.
- **Block 3C** looks at how forces can deform solid materials. It shows how to calculate stresses and strains and interpret stress–strain graphs in order to predict how materials will behave when forces are applied.

Block 3A: Electric circuits, pages 62–73

Ideas from GCSE	Content outline of Block 3A
<ul><li>Relationship between current, voltage and resistance</li><li>Measuring current and voltage</li><li>Nature of electric current</li><li>Energy and power in electric circuits</li></ul>	<ul><li>Electric current and potential difference</li><li>Resistances in series and parallel</li><li>Ohm's law</li><li>Resistivity</li><li>Electrical energy and power</li><li>Kirchhoff's laws</li><li>Electromotive force, internal resistance and potential dividers</li></ul>

Block 3B: Alternating current, pages 74–77

Ideas from GCSE	Content outline of Block 3B
	<ul><li>Peak, peak-to-peak and r.m.s. values</li><li>Using an oscilloscope</li></ul>

Block 3C: Elastic properties of solids, pages 78–81

Ideas from GCSE	Content outline of Block 3C
<ul><li>Stretching effect of a force</li><li>Hooke's law</li></ul>	<ul><li>Hooke's law and elastic limit</li><li>Elastic strain energy</li><li>Stress and strain</li><li>Young modulus</li></ul>

End-of-module questions, pages 82–83

Electric current and voltage

A battery pushes charge around a circuit. A current is considered to be a flow of *positive* charge. Because of this, current must flow from positive to negative. This is known as **conventional current**.

The charged particles in metals are **electrons**. Electrons are negatively charged, so they flow from negative to positive. We still say that the current flows from positive to negative.

Coulombs and amps

Charge Q is measured in **coulombs** (C).

Current I is measured in amperes, **amps** (A).

One amp is one coulomb per second: $1\ A = 1\ C\ s^{-1}$.

One coulomb is the charge that passes when a current of 1 amp flows for 1 second.

Formulae relating Q, I and t

To calculate the current I flowing when charge ΔQ passes a point in time interval Δt:

$$I = \frac{\Delta Q}{\Delta t}$$

This reminds us that current is a *rate* of flow – that's why we divide by time.

To calculate the charge:

$$\Delta Q = I \times \Delta t$$

The bigger the current, and the longer it flows, the more the charge that passes.

Worked examples

1 What current flows when a charge of 600 C passes a point in 1 minute?

    Step 1 Write down what you know, and what you want to know:

$$\Delta Q = 600\ C, \quad \Delta t = 60\ s, \quad I = ?$$

    Step 2 Choose the appropriate equation, substitute and solve:

$$I = \frac{\Delta Q}{\Delta t} = \frac{600\ C}{60\ s} = 10\ A$$

> ▶ Use SI units.

> ▶ Don't forget units. Strictly speaking, this is the *average* current flowing in this time.

2 How much charge passes a point when a current of 10 mA flows for 10 s?

Step 1 Write down what you know, and what you want to know:

$$I = 10 \text{ mA (or } 10^{-2} \text{ A or } 0.01 \text{ A)}, \quad \Delta t = 10 \text{ s}, \quad \Delta Q = ?$$

Step 2 Choose the appropriate equation, substitute and solve:

$$\Delta Q = I \times \Delta t$$
$$= 10 \text{ mA} \times 10 \text{ s} = 100 \text{ mC}$$

> Since we worked in mA (milliamps), the answer is in mC (millicoulombs). We could have converted from mA to A.

> ✓ *Quick check 1, 2*

Voltage or potential difference

A **voltage** *across* a component will cause a current to flow *through* the component. When current flows, electrical energy is converted into other forms. The voltage between two points is a measure of how much energy is converted for each coulomb of charge that flows from one point to the other.

Potential difference (p.d.) is simply another (more correct) term for voltage.

The voltage or p.d. *V* across a component, such as a resistor or lamp, tells you how many joules of energy *W* are converted for each coulomb of charge *Q* passing through:

$$V = \frac{W}{Q}$$

Voltage or potential difference *V* is measured in volts (V).

Energy *W* is measured in joules (J).

> Think of *W* for work. Avoid using *E* for energy.

Charge *Q* is measured in coulombs (C).

One volt is one joule per coulomb: $1 \text{ V} = 1 \text{ J C}^{-1}$.

> ✓ *Quick check 3–6*

? *Quick check questions*

1 What current is flowing if 240 mC of charge flows past a point in 30 s?

2 A motorist is having trouble getting a car to start. The battery supplies a current of 100 A for 1 minute. How much charge flows from the battery in this time?

3 How much energy is transferred to each coulomb of charge by a 9 V battery?

4 What is the p.d. across a resistor in which 100 J of energy is converted to heat for every 2.4 C that flows through it?

5 10 C of charge flows through a p.d. of 6 V. How much energy is transferred?

6 A current of 2.5 A flows through a resistor for 1 minute. It transfers 600 J of energy to the resistor. What is the p.d. across the resistor?

> Start by calculating how much charge flows.

Resistance

A **potential difference** (p.d.) is needed to push a current through a component. The **electrical resistance** of the component tells us how easy (or rather, how difficult) it is to make current flow through it.

> **The greater the resistance, the smaller the current that flows for a given p.d.**

▸▸ *For the meaning of potential difference (p.d.) see page 63.*

Defining resistance

The **resistance** (R) of a component is the ratio of the p.d. (V) *across* it to the current (I) flowing *through* it. It is defined by the equation:

$$\text{resistance} = \frac{\text{p.d.}}{\text{current}} \qquad R = \frac{V}{I}$$

The equation for resistance can be rearranged:

$$V = IR \qquad I = \frac{V}{R}$$

Ohms, amps and volts

Resistance is measured in **ohms** (Ω). One ohm is one volt per amp: $\mathbf{1\ \Omega = 1\ V\ A^{-1}}$.

So it takes a p.d. of 1 V to make a current of 1 A flow through a 1 Ω resistor, and it takes a p.d. of 10 V to make a current of 1 A flow through a 10 Ω resistor.

- 1 kilohm = 1 kΩ = 10^3 Ω = 1000 Ω
- 1 megohm = 1 MΩ = 10^6 Ω = 1000 000 Ω

> ❶ The symbol Ω is the Greek letter omega.

> ✓ *Quick check 1, 2*

Resistors in series

When resistors are connected **in series** (end-to-end), the current flows through one and then through the next, and so on.

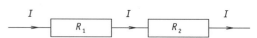

- Resistors in series must have the *same current* flowing through them.
- The p.d. of the supply is *shared* between them.

To find the combined resistance R of two or more resistors in series, add up their individual resistances:

> $$R = R_1 + R_2 + R_3 + \text{... in series}$$

Worked example

A 20 Ω resistor and a 5 Ω resistor are connected in series with a 10 V battery. What is the p.d. across each resistor?

Step 1 Sketch a diagram, and mark on it the available information.
Step 2 Calculate the combined resistance:

$$R = R_1 + R_2 = 20\ \Omega + 5\ \Omega = 25\ \Omega$$

Step 3 Calculate the current that flows:

$$I = \frac{V}{R} = \frac{10\ V}{25\ \Omega} = 0.4\ A$$

Step 4 Calculate the p.d. across each resistor:

Across 20 Ω: $V = IR = 0.4\ A \times 20\ \Omega = 8\ V$
Across 5 Ω: $V = IR = 0.4\ A \times 5\ \Omega = 2\ V$

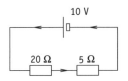

▶ A useful rule: The bigger resistor gets a bigger share of the p.d.

✓ *Quick check 3*

Resistors in parallel

When resistors are connected **in parallel** (side-by-side), the *current* divides up, part of it flowing through each resistor.

- Resistors in parallel have the *same p.d.* across them.
- The current flowing from the supply is *shared* between them.

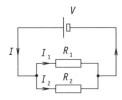

Since $I = V/R$ for *each* resistor (see opposite), to find the combined resistance R of two or more resistors in series, add up the *reciprocals* of their individual resistances:

$$\frac{1}{R} = \frac{1}{R_1} + \frac{1}{R_2} + \frac{1}{R_3} + \dots \text{ in parallel}$$

Worked example

A 20 Ω resistor and a 5 Ω resistor are connected in parallel with a 10 V battery. What current flows from the battery?

Step 1 Sketch a diagram, and mark on it the available information.
Step 2 Calculate (in two stages!) the combined resistance:

$$\frac{1}{R} = \frac{1}{R_1} + \frac{1}{R_2} = \frac{1}{20\ \Omega} + \frac{1}{5\ \Omega} = 0.05\ \Omega^{-1} + 0.20\ \Omega^{-1} = 0.25\ \Omega^{-1}$$

$$R = \frac{1}{0.25\ \Omega^{-1}} = 4\ \Omega \quad (R \text{ is always less than the smallest of } R_1, R_2, \text{ etc.})$$

Step 3 Calculate the current from the combined resistance and the p.d.:

$$I = \frac{V}{R} = \frac{10\ V}{4\ \Omega} = 2.5\ A$$

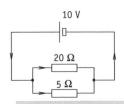

Another method: Calculate the current through each resistor separately, and add them together.

✓ *Quick check 4, 5*

❓ **Quick check questions**

1 What is the resistance of a resistor if a p.d. of 10 V makes a current of 2 A flow through it?

2 What p.d. will make a current of 20 mA flow through a 500 kΩ resistor?

3 Two 20 Ω resistors are connected in series with a 5 V supply. What is the p.d. across each resistor?

4 What is the resistance of 20 Ω, 30 Ω, 60 Ω resistors connected in parallel?

5 Three 2 kΩ resistors are connected in parallel across a 12 V supply. What current flows from the supply?

▶ Recall that an ohm is a volt per amp.

▶ How is the p.d. shared between resistors in series?

▶ First calculate the current that flows through one resistor.

Ohm's law

A **potential difference** (p.d.) or **voltage** pushes current through a conductor. The greater the p.d., the greater the current. An **ohmic conductor** is one in which the current flowing is proportional to the p.d. pushing it. It obeys **Ohm's law**.

Remembering the equation $V = IR$ (page 64), this means that the resistance of an ohmic conductor is constant, and does not depend on the p.d. across it.

Measuring resistance

- An **ammeter** (or datalogging current sensor) is an instrument for measuring current, for example through a resistor. It is connected *in series* with the resistor.

- A **voltmeter** (or datalogging voltage sensor) is an instrument for measuring the p.d. (voltage) across a resistor. It is connected *in parallel* with the resistor.

- Reversing the connections to the resistor makes the current flow through it in the opposite direction. This gives negative values of current I and voltage V.

- The results can be plotted as a **current–voltage characteristic** graph, or I–V graph. The figure shows an I–V graph for a metallic conductor at constant temperature. An ohmic conductor gives a straight line through the origin.

It is usual to plot p.d. on the x-axis.

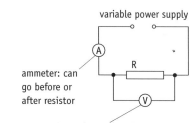

variable power supply

ammeter: can go before or after resistor

voltmeter: usually connected last

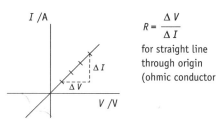

$$R = \frac{\Delta V}{\Delta I}$$

for straight line through origin (ohmic conductor

✓ *Quick check 1*

Non-ohmic conductors

Metals obey Ohm's law – they are ohmic conductors – provided the temperature remains constant. The I–V characteristic graph of a **non-ohmic conductor** is not a straight line. The figure shows two examples.

- *Filament lamp:* As the metal filament gets hotter with increasing current and voltage, its resistance increases. The current is less than it would be if it remained proportional to the voltage.

- *Semiconductor diode:* In the forward direction, this allows current to flow when the p.d. is above 0.7 V (for a silicon diode). In the reverse direction, only a very small current flows.

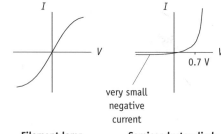

very small negative current

Filament lamp **Semiconductor diode**

✓ *Quick check 2*

Temperature dependence

- *Metal:* The resistance increases gradually as the temperature is increased. (Atomic vibrations increase, so conduction electrons are scattered more.)

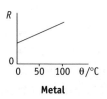

Metal

- **NTC thermistor:** The resistance decreases rapidly over a narrow range of temperature.

NTC means 'negative temperature coefficient': hotter = less resistance.

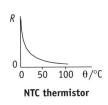

NTC thermistor

Resistivity

Some materials resist the flow of electric current more than others. The property that describes this is **resistivity**, ρ (Greek letter rho).

To calculate the resistance R of a wire, for example, we need to know three things:

- its length l – the longer the wire, the greater its resistance: $R \propto l$

- its cross-sectional area A – the fatter the wire, the less its resistance: $R \propto 1/A$

- the resistivity of the material ρ : $R \propto \rho$ (ρ is a constant for a given material, at a given temperature). The formula is

material of resistivity ρ

$$R = \frac{\rho l}{A} \quad \text{or} \quad \rho = \frac{RA}{l}$$

Units Resistivity ρ is measured in Ω m (**ohm metres**).

A typical value for a good conductor: resistivity of copper = 1.7×10^{-8} Ω m.

▶ Take care – not ohms per metre!

Worked example

What is the resistance of a 20 m length of silver wire of diameter 1.0 mm? (Resistivity of silver = 1.6×10^{-8} Ω m.)

Step 1 Write down what you know, and what you want to know; you will have to calculate A.

$$l = 20 \text{ m}, \quad \rho = 1.6 \times 10^{-8} \text{ } \Omega \text{ m}$$
$$A = \pi r^2 = \pi \times (0.5 \times 10^{-3} \text{ m})^2 = 7.9 \times 10^{-7} \text{ m}^2, \quad R = ?$$

Step 2 Calculate R.

$$R = \frac{\rho l}{A} = \frac{1.6 \times 10^{-8} \text{ } \Omega \text{ m} \times 20 \text{ m}}{7.9 \times 10^{-7} \text{ m}^2} = 0.41 \text{ } \Omega$$

▶ Remember to halve the diameter to find the radius.

✓ **Quick check 3, 4**

? *Quick check questions*

1 The table shows experimental results for a carbon resistor of resistance R. Plot a graph and use it to deduce R.

current I/mA	180	340	550	700	910
p.d. V/V	2.0	4.0	6.0	8.0	10.0

2 Use the graph in the figure to decide at what voltage the filament lamp and the resistor have the same resistance R. What is the value of R?

3 What is the resistance of a 5.0 m length of copper wire of cross-sectional area 1.0 mm²? (Resistivity of copper = 1.7×10^{-8} Ω m.)

4 Does the resistivity of a metal increase or decrease with temperature?

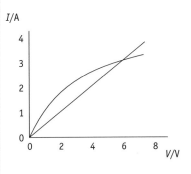

▶ 1 mm² = 10^{-6} m².

Electrical power

We use electricity as a convenient way of transferring energy from place to place. *Voltage* is a measure of how much energy is transferred to or from each coulomb of charge. *Current* tells us about the rate at which charge moves. Combining these quantities tells us about the *rate* at which energy is transferred by a current – the **electrical power**.

Power in general

We met the general idea of power on page 47. Power is the rate at which energy is transferred.

$$\text{power} = \frac{\text{energy transferred}}{\text{time taken}} \quad P = \frac{W}{t}$$

$$\text{energy transferred} = \text{power} \times \text{time} \quad W = Pt$$

Units

Power is measured in **watts**, W.

1 watt = 1 joule per second $1\ \text{W} = 1\ \text{J s}^{-1}$

1 kilowatt = 1 kW = 10^3 W = 1000 W

1 megawatt = 1 MW = 10^6 W = 1000 000 W

1 gigawatt = 1 GW = 10^9 W = 1000 000 000 W

▶▶ *Mechanical power is the rate at which energy is transferred by a force – see page 47.*

✓ *Quick check 1, 2*

Calculating electrical power

The greater the current I and the greater the p.d. V that it is flowing through, the greater the power P.

$$\text{power} = \text{current} \times \text{p.d.} \quad P = IV$$

Combining $W = Pt$ with $P = IV$ gives

$$W = IVt$$

Combining $P = IV$ with $V = IR$ gives two alternative forms:

$$P = I^2R \quad \text{and} \quad P = \frac{V^2}{R}$$

The equation $P = I^2R$ helps to explain why we use high voltage and low current when transmitting electrical power over large distances. For a given power P, the higher the voltage, the lower the current. A low value of I is important because the power loss in the cables is I^2R, e.g. doubling the current gives four times the power loss in the cables.

🔘 You may find this easier to remember as watts = amps × volts.

✓ *Quick check 3*

🔘 Choose the form of equation according to the information you have in any question, e.g. given V and R, use $P = V^2/R$.

Worked example

A car headlamp bulb is labelled '12 V, 48 W'. What is its resistance in normal operation? (The label indicates its normal operating voltage and power.)

Step 1   Write down what you know, and what you want to know:

$$V = 12 \text{ V}, \quad P = 48 \text{ W}, \quad R = ?$$

Step 2   Select the appropriate equation and rearrange it to make R the subject:

$$P = \frac{V^2}{R} \quad \text{so} \quad R = \frac{V^2}{P}$$

Step 3   Substitute values and calculate the answer:

$$R = \frac{(12 \text{ V})^2}{48 \text{ W}} = 3 \text{ } \Omega$$

✓ Quick check 4

SI unit summary

It's often easier to learn equations linking units rather than quantities.

quantity	unit	equivalents	in words
current I	ampere, A	$1 \text{ A} = 1 \text{ C s}^{-1}$	amp = coulomb per second
p.d. V	volt, V	$1 \text{ V} = 1 \text{ J C}^{-1}$	volt = joule per coulomb
resistance R	ohm, Ω	$1 \text{ } \Omega = 1 \text{ V A}^{-1}$	ohm = volt per amp
power P	watt, W	$1 \text{ W} = 1 \text{ J s}^{-1}$	watt = joule per second

? Quick check questions

1   The output of a power station is stated as 450 MW. How many joules of electrical energy does it supply each second?

2   A battery transfers 30 J of energy each minute. What power is this?

3   An electric motor draws a current of 2.5 A from a 12 V supply. What power does it transfer?

4   A 60 W lamp has a resistance of 2.4 Ω. What current flows through it in normal operation?

5   Explain why the cables connecting the battery to the starter motor of a car engine are much thicker than the cables for the other electrical components such as the lights.

Kirchhoff's laws

These two laws summarise the application of two important principles to d.c. circuits – the *conservation of charge or current* and the *conservation of energy*.

Conservation of current: Kirchhoff's first law

Current is the *flow of charge*. Charge cannot disappear or get used up. For this reason, we say that **charge and current are conserved**.

- At point X, current splits up: $I = I_1 + I_2$
- At point Y, currents recombine: $I_1 + I_2 = I$

These equations describe the conservation of current. They are an example of **Kirchhoff's first law**. A more formal statement is:

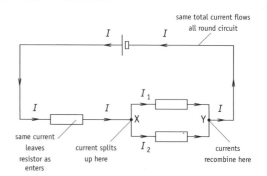

same total current flows all round circuit

same current leaves resistor as enters

current splits up here

currents recombine here

> **The sum of the currents entering a point is equal to the sum of the currents leaving the point.**
> $\Sigma I_{in} = \Sigma I_{out}$ **where Σ (sigma) means 'the sum of'.**

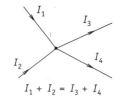

$I_1 + I_2 = I_3 + I_4$

✓ *Quick check 1, 2*

p.d. and e.m.f.

We use electricity to transfer energy from place to place. If you use a battery to light a bulb, electricity transfers energy from the battery to the bulb. The *voltage* of a source of electricity is a measure of these energy transfers.

We use two (more correct) terms for voltage:

- **potential difference**, p.d., symbol V
- **electromotive force**, e.m.f., symbol E

> ► Take care always to talk about the potential difference *across* a component, or *between* two points. A p.d. does not 'go through' a component.

Around a circuit

In this circuit, the cell is pushing a current (a flow of charge) through the lamp.

The current is the same at all points around the circuit.

- *Inside the cell:* Charges flow through the cell, collecting energy.
- *Inside the lamp:* The same charges flow through the lamp, giving up energy.

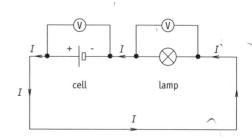

cell

lamp

The voltmeters give equal but opposite readings (positive and negative).

Note that the current inside the cell flows from negative to positive – see page 62.

The meaning of e.m.f.

The voltage shown on a cell or battery tells you its **e.m.f.** From this, you can tell the amount of energy given to each coulomb of charge that passes through.

> **The e.m.f. of a supply is the work it does in pushing 1 C of charge around a complete circuit.**

- A 1.5 V cell gives 1.5 J to each coulomb.
- A 6 V battery gives 6 J to each coulomb.
- The 230 V mains gives 230 J to each coulomb.

> **Similarly, the p.d. across a component (such as a resistor or lamp) tells you how many joules of energy are given up by each coulomb passing through.**

▶▶ *For more information on measuring e.m.f. see page 72.*

Conservation of energy: Kirchhoff's second law

Trace the movement of 1 C of charge around the circuit.

- At **A**: 5 J gained
- At **B**: 15 J gained; total energy gained = 20 J
- At **C**: 10 J lost
- At **D**: 10 J lost; total energy lost = 20 J

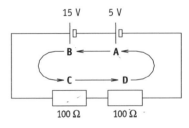

By the time the charge has completed the circuit, it has lost as much energy as it gained. This is an example of **conservation of energy**. We can think of this in terms of voltages around the circuit:

> *Kirchhoff's second law:*
> **The sum of the e.m.f.s around any circuit loop is equal to the sum of the p.d.s: $\Sigma E = \Sigma IR$.**

✓ *Quick check 3–5*

? *Quick check questions*

1 Calculate current I_4, as shown in the figure.

2 Currents of 2.5 A, 1.0 A and 10.5 A are supplied by a car battery. What is the total current it supplies?

3 Two resistors are connected in series with the 230 V mains supply. The p.d. across one resistor is 70 V. What is the p.d. across the other?

4 For the circuit shown, calculate:

  a the total e.m.f.;

  b the current that flows;

  c the p.d. across each resistor.

  Show that Kirchhoff's second law is satisfied.

5 How much work is done by the 230 V mains supply in pushing 1 C of charge round a circuit?

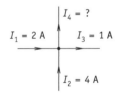

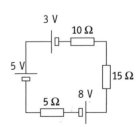

Internal resistance and potential dividers

We use a cell or power supply to provide a p.d. Sometimes it provides less than we expect, because of its **internal resistance**. Sometimes we want a smaller p.d. than it supplies, and we use a **potential divider** to reduce the p.d.

Internal resistance

Think about the current flowing around a circuit, pushed by a cell or power supply. The same current flows all the way round. In particular, it flows through the *interior* of the supply (from negative to positive). The interior of a supply is made up of chemicals or metal wire, and must have resistance. This is the internal resistance r of the supply.

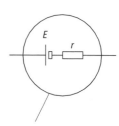

We show the cell as a 'perfect cell', marked with its e.m.f. E, and a small resistor r in series with it. The current I flows through a combined resistance $R + r$.

$$E = I (R + r)$$

$$E - Ir = IR$$

the circle is optional; it indicates that E and r are part of the same thing

e.m.f. – 'lost' voltage = terminal p.d. of cell.

We get fewer volts out of the supply because some of the e.m.f. is used up in overcoming the internal resistance when a current is flowing.

▶▶ *The meaning of e.m.f. is explained on page 70.*

> ● You can think of the voltage across the internal resistance as 'lost' volts.

Measuring E and r

Varying the value of R makes the current I change. The graph shows that the greater the current that flows from the supply, the smaller its terminal p.d. The graph is roughly a straight line:

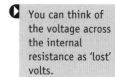

- gradient = $-r$,
- intercept on y-axis = E.

variable resistor

Using a high-resistance voltmeter

A digital voltmeter, or a datalogging voltage sensor, has a resistance of millions of ohms. Connect it across a supply, and only a tiny (negligible) current will flow. Its reading will therefore indicate the e.m.f. of the supply ('lost volts' = 0).

> **The e.m.f. of a supply is the p.d. across its terminals when no current flows. The supply is said to be 'open circuit'.**

✓ *Quick check 1, 2*

Worked example

A power supply of e.m.f. 6 V and internal resistance 2 Ω is connected across a 10 Ω resistor. What current flows through the resistor, and what is the terminal p.d. of the supply?

Step 1 Draw a diagram, showing both R and r.
Step 2 Calculate the total resistance in the circuit.

$$R + r = 10\ \Omega + 2\ \Omega = 12\ \Omega$$

Step 3 Calculate the current that flows.

$$I = \frac{E}{R + r} = \frac{6\ V}{12\ \Omega} = 0.5\ A$$

Step 4 Calculate the terminal p.d.

$$V = IR = 0.5\ A \times 10\ \Omega = 5\ V$$

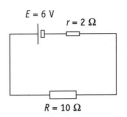

$E = 6\ V$ $r = 2\ \Omega$

$R = 10\ \Omega$

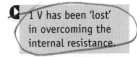

1 V has been 'lost' in overcoming the internal resistance.

✓ **Quick check 3**

Potential dividers

Reduce the p.d. provided by a supply by connecting two resistors across its terminals. Tap off the required p.d. V_{out} from the point between them.

The bigger resistance takes the bigger share of the p.d.:

$$\frac{V_1}{V_2} = \frac{R_1}{R_2}$$

Use equal resistances to give *half* the p.d. of the supply.

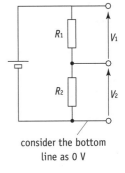

R_1 V_1

R_2 V_2

consider the bottom line as 0 V

Varying resistances

- Shining light on the LDR (light-dependent resistor) will decrease its resistance. V_{out} will decrease.

- Warming the NTC thermistor will decrease its resistance. The p.d. across the thermistor will decrease. So V_{out} will increase.

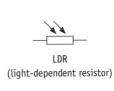

LDR
(light-dependent resistor)

NTC thermistor

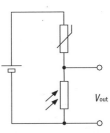

V_{out}

✓ **Quick check 3, 4**

? Quick check questions

1 A power supply of e.m.f. 500 V and internal resistance 0.1 Ω is connected to a heater of resistance 125 Ω. Calculate the current that flows through the heater.

2 A high-resistance voltmeter is connected across a cell, and gives a reading of 1.55 V. A 600 Ω resistor is added in parallel with the voltmeter, which now reads 1.50 V. What are the e.m.f. and internal resistance of the cell?

3 Calculate V_{out} as shown in the figure. Notice that one resistor has twice the resistance of the other.

4 A potential divider is constructed from a 50 Ω fixed resistor and a thermistor whose resistance changes from 450 Ω at 20 °C to 50 Ω at 80 °C. The divider is connected across a 10 V supply. Draw a diagram of this arrangement, and show that the voltage across the thermistor will vary between 9 V and 5 V as the temperature is varied between 20 °C and 80 °C.

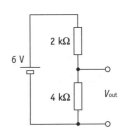

$2\ k\Omega$

6 V

$4\ k\Omega$ V_{out}

Alternating currents

One of the biggest advantages of using *alternating current* (*a.c.*) is that voltages can be easily stepped up or down using transformers, making for much greater efficiency in the distribution of electricity.

- In **direct current** (**d.c.**) electrons travel in a conductor in *one direction only*.

- In **alternating current** (**a.c.**) the electrons move *back and forth*.

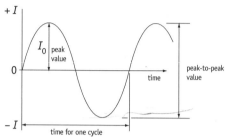

One complete back-and-forth movement is **one cycle**. The most common type of a.c. is **sinusoidal** – this means it has a sine wave shape.

The current is changing all the time – how should we measure it? The *average* value of I (or V) over one cycle is zero, so this is not very useful.

We could use the **peak value** (from zero to a 'peak' or 'trough' of the wave shape). Peak current is given the symbol I_0. It is the amplitude of the wave shape. Peak voltage is V_0.

Or we could use the **peak-to-peak value**, which is from positive peak to negative peak.

Root mean square values

The value most often used to describe or measure a.c. is called the **root mean square** value, abbreviated to **r.m.s.**

Both a.c. and d.c. light up a filament lamp. They both produce heating in a resistor.

The r.m.s. value of an a.c. current is the value of steady direct current that gives the same heating or power in a resistor as the a.c.

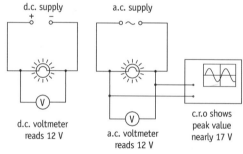

If a light bulb lights at normal brightness when connected to a 12 V d.c. supply, such as a car battery, it will also light at normal brightness when connected to an a.c. **r.m.s. voltage** of 12 V. The same heating would be produced in the filament. If this a.c. voltage is displayed on a cathode ray oscilloscope (see page 76), the *peak* value is about 17 V.

The heating in a resistor is given by $P = I^2R$ (see page 68).

> ▶ Remember: the r.m.s. a.c. current is the same as the d.c. current.

If a.c. is used, I changes all the time, so we square all the currents over one cycle and find the mean or average. The diagram shows that the average value of I^2 over a cycle is $I_0^2/2$.

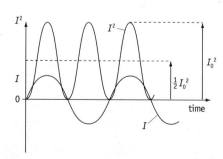

Equating r.m.s. a.c. power to average d.c. power,

$$I_{rms}^2 R = \left(\frac{I_0^2}{2}\right) \times R$$

Cancelling R and taking the square root, we get:

$$I_{rms} = \sqrt{\frac{I_0^2}{2}} = \frac{I_0}{\sqrt{2}}$$

Applying a similar analysis to $P = V^2/R$, we get

$$V_{rms} = \frac{V_0}{\sqrt{2}}$$

Summarising:

r.m.s. value = peak value/√2	**peak value = √2 × r.m.s. value**

$$I_{rms} = \frac{I_0}{\sqrt{2}} = 0.71 I_0 \qquad I_0 = \sqrt{2} \times I_{rms} = 1.41 I_{rms}$$

$$V_{rms} = \frac{V_0}{\sqrt{2}} = 0.71 V_0 \qquad V_0 = \sqrt{2} \times V_{rms} = 1.41 V_{rms}$$

a.c. ammeters and voltmeters are calibrated to give r.m.s. currents and voltages. Always assume that a.c. voltages and currents are r.m.s. values unless told otherwise.

Worked examples

1 The r.m.s. voltage of the a.c. mains is 230 V. What is the peak value of the a.c. mains voltage?

$$V_0 = \sqrt{2} \times V_{rms} = \sqrt{2} \times 230\ V = 325\ V$$

2 If a hair dryer rated 230 V, 1600 W is connected to the mains and switched on, what r.m.s. current will it take?

$$P = I \times V$$

$$I = \frac{P}{V} = \frac{1600\ W}{230\ V} = 7.0\ A$$

> The insulation of a mains-operated appliance must be able to withstand this voltage.

> ✓ *Quick check 1,2*

> By using r.m.s. values, we can solve a.c. problems using the same formulae as we use for d.c. It is not necessary to write V_{rms} or I_{rms}. Write V and I as usual.

> ✓ *Quick check 3*

❓ Quick check questions

1 The peak voltage of a sinusoidal a.c. is 22 V. What is

  a the peak-to-peak voltage,

  b the r.m.s. voltage?

2 The r.m.s. current in a 20 Ω resistor is 25 mA. Calculate the peak current in the resistor and the peak voltage across it.

3 A laboratory electric heater takes an r.m.s. current of 4.2 A from a 12 V a.c. supply. Calculate the power of the heater. If the heater is to provide the same power when connected to a 12 V battery, what current will it take?

The cathode ray oscilloscope (c.r.o.)

The **cathode ray oscilloscope (c.r.o.)** is an instrument for measuring voltages, time intervals and frequencies, and for showing the shapes of waves.

To start the instrument, switch it on, turn timebase or SWP off, adjust ↔ or ↕ controls to centralise the spot, and adjust focus and intensity. Then connect the signal to be measured or displayed to the Y input.

To measure d.c. voltage

Step 1 Connect voltage to be measured to Y input. The spot will move up or down the screen a distance y.

Step 2 Find unknown voltage by
$$V = y \times \text{volt/div}$$

$V = 2.5 \text{ div} \times 0.5 \text{ V/div}$
$= 1.25 \text{ V}$

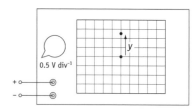

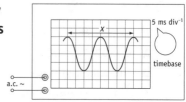
Alter the Y amplifier setting if the spot hardly moves, or if it moves off screen.

To measure a.c. voltage

Step 1 Connect voltage to be measured to Y input.

Step 2 Adjust Y amplifier so that line fills most of screen height.

Y amp setting = 5 V/div

Step 3 Measure y, the line length. Calculate peak voltage V_0 by

$y = 4.6 \text{ div}$

$$V_0 = \text{volt/div} \times y/2$$
$$V_{rms} = 0.71 \times V_0 \text{ if a.c. is a sine wave}$$

$V_0 = 5 \text{ V/div} \times 2.3 = 11.5 \text{ V}$
$V_{rms} = 0.71 \times 11.5 \text{ V} = 8.17 \text{ V}$

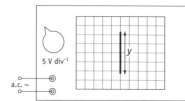

To measure frequency

Step 1 Connect signal to Y input.

Step 2 Switch on timebase. Set variable control to 'cal'. Adjust timebase to give 2 or 3 cycles on screen.

Timebase setting = 5 ms/div
2 waves occupy 8.0 divisions

Step 3 Measure width x for a number of complete cycles.

$x = 8.0 \text{ div}$

Step 4 Calculate time represented by this number of divisions.

time = 8.0 div × 5.0 ms/div = 40 ms

Step 5 Calculate time period:
   T = total time/number of cycles

$T = 40/2 = 20$ ms $= 0.02$ s

Step 6 Calculate frequency:
   f = 1/time period

$f = 1/T = 1/0.02 = 50$ Hz

✓ Quick check 1

To measure a time interval between two signals

Step 1 Apply signals to Y input.

Step 2 Switch on timebase. Set variable control to 'cal'. Adjust timebase to give several divisions between signals.

Timebase setting = 20 ms/div

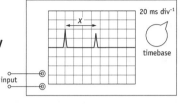

Step 3 Measure x.

x = **3.5 divisions**

Step 4 Time interval = $x \times$ timebase setting.

**3.5 div × 20 ms/div = 70 ms
= 0.07 s**

✓ Quick check 2

? Quick check questions

1 The waveform shown in the figure was displayed on a c.r.o. with the Y amplifier set at 0.5 V/cm and the timebase at 50 ms/cm.

Determine

   a the peak voltage,

   b the r.m.s. voltage,

   c the frequency.

2 A c.r.o. was used to measure the time interval between an initial pulse of ultrasound passed into a piece of steel and the pulse reflected from a crack inside the steel. The timebase was set at 0.5 μs/cm and the separation of the pulses on the screen was measured as 4.2 cm. What was the time interval between the pulses?

Deforming solids

Some properties of materials are more important than others when selecting a
suitable material for a particular application. An aircraft wing must have high
strength but low weight; a drill bit must be strong and hard, but its weight is of
little importance.

Density

Density is defined as mass m per unit volume V. The symbol for density is ρ (Greek
letter *rho*):

$$\rho = \frac{m}{V}$$

The units of density are kg m^{-3}.

✓ Quick check 1

Hooke's law

The greater the **load** (the force stretching the spring), the
greater its **extension** (increase in length). Eventually the load
is so great that the spring becomes permanently stretched. The
graph shows two things:

- At first the graph is a straight line, so load F is proportional
  to extension e:

$$F \propto e$$

This is **Hooke's law**.

- Beyond the **elastic limit**, the spring does not return to its original length when
  the load is removed.

✓ Quick check 2

Elastic strain energy

Work has to be done to stretch a spring or wire. The work done is equal to
the elastic strain energy stored in the stretched spring or wire.

If the material is not taken beyond the Hooke's law region:

- work done when force F causes extension e = average force × extension,
- average force × extension = area under force–extension graph = $\frac{1}{2}Fe$.

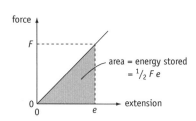

area = energy stored
= $^{1}/_{2}\,F\,e$

Hence

$$\text{strain energy} = \tfrac{1}{2}Fe$$

If the material is taken to a point *beyond* the Hooke's law region, the strain energy
is the area under the force–extension graph up to that point.

Stress and strain

If you were to compare the results of stretching a short thick nylon rod with a long thin steel wire, the nylon rod might break at a higher force than the steel wire, and it may not show as much extension. It would not be fair to say that nylon is stronger than steel, or that steel stretches more easily.

In order to compare one material with another, we must take into account the original length and cross-sectional area of the specimen of material under test. We do this by using **stress** and **strain** as follows:

$$\text{stress} = \frac{\text{load}}{\text{original cross-sectional area}} \quad \sigma = \frac{F}{A} \quad \textbf{(units N m}^{-2} \textbf{ or Pa, pascal)}$$

$$\text{strain} = \frac{\text{extension}}{\text{original length}} \quad \varepsilon = \frac{e}{l} \quad \textbf{(strain has no units)}$$

> ▶ For metals, stresses are usually large (e.g. MPa) and strains are usually small ($< 10^{-2}$). Use this as a check on your answers.

> ▶ σ is pronounced 'sigma', and ε is pronounced 'epsilon'.

Stress–strain graphs

At low stresses, below the elastic limit, the material will return to its original length when the load is removed. This is **elastic deformation**. At higher stresses, the material becomes permanently deformed. This is **plastic deformation**.

Once the stress reaches its highest value, called the **ultimate tensile stress** (u.t.s.) or the **tensile strength** or the **breaking stress**, the material will break.

Materials may be classified as follows:

- A **ductile** material, such as copper, stretches a lot beyond the elastic limit. The specimen shows 'necking' or deformation before fracture.

- A **brittle** material, such as cast iron or glass, snaps suddenly, showing little or no deformation before fracture. The broken pieces fit together.

- A **polymeric** material, such as polythene, does not show linear behaviour and gives a high strain at relatively low stress.

> ✓ *Quick check 3, 4*

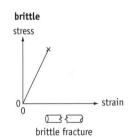

ductile fracture

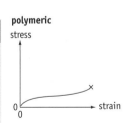

brittle fracture

polymeric

stress ↗ strain

> ▶ Don't forget to change mm² to m².

❓ *Quick check questions*

1. What is the mass of a concrete garage base measuring 5.0 m × 3.1 m × 0.18 m? The density of concrete is 2400 kg m^{-3}.

2. A spring is 1.2 m long when unstretched, and 1.4 m long when a load of 50 N is applied. Its elastic limit is 80 N. Calculate the extension produced by the 50 N load.

3. Calculate the stress in a steel wire of cross-sectional area 0.25 mm^2 under a tension force of 60 N.

4. Calculate the strain in a strip of rubber of initial length 22 cm when extended by 7.2 cm under a tensile force.

The Young modulus

The Young modulus is a property of a material that tells us how stiff or resistant it is to stretching. It is a very useful property because it links stress and strain, and gives an indication of how inflexible a material is.

For materials whose stress–strain graph is initially a straight line,

$$\textbf{stress} \propto \textbf{strain} \quad \text{i.e.} \quad \frac{\textbf{stress}}{\textbf{strain}} = \textbf{constant}$$

This constant is known as the **Young modulus**, E, of the material. In terms of the quantities introduced on pages 78–79,

$$E = \frac{\textbf{stress}}{\textbf{strain}} = \frac{\sigma}{\varepsilon} = \frac{F/A}{e/l} = \frac{Fl}{Ae}$$

stress

Young modulus = slope

E is measured in N m^{-2} or pascals (Pa). Because values of E are usually large, typical values are expressed in MPa ($\times 10^6$ Pa) or even GPa ($\times 10^9$ Pa).

E is the slope of the initial part of the stress–strain graph. The steeper the slope, the stiffer or more resistant to stretching the material is.

We have seen (page 78) that the area under a force–extension graph is *strain energy*. The area under a stress–strain graph is the *strain energy per unit volume* (in J m^{-3}).

Worked example

A lift weighing 2.00×10^4 N is suspended by a 15.0 m long steel cable of cross-sectional area 1.40×10^{-4} m^2. Determine the stress in the cable, and the extension of the cable. The Young modulus for steel is 210 GPa.

Step 1 Write down what you know, and what you want to know:

$$F = 2.00 \times 10^4 \text{ N}, \quad A = 1.40 \times 10^{-4} \text{ m}^2, \quad E = 210 \times 10^9 \text{ Pa}$$
$$l = 15.0 \text{ m}, \quad \sigma = ?, \quad e = ?$$

Step 2 Calculate the stress:

$$\sigma = \frac{F}{A} = \frac{2.00 \times 10^4 \text{ N}}{1.40 \times 10^{-4} \text{ m}^2} = 142 \text{ MPa}$$

Step 3 To find the extension, rearrange $E = \dfrac{\sigma}{\varepsilon}$ to get

$$\varepsilon = \frac{\sigma}{E} = \frac{142 \times 10^6 \text{ Pa}}{210 \times 10^9 \text{ Pa}} = 6.80 \times 10^{-4}$$

Step 4 Finally, rearrange $\varepsilon = \dfrac{e}{l}$ to get the extension:

$$e = \varepsilon \times l = 6.80 \times 10^{-4} \times 15.0 \text{ m} = 0.01 \text{ m}$$

✓ *Quick check 1,2*

Measuring the Young modulus

- Use a *long* wire – it will give a greater extension for a given load, so less uncertainty in measurement.

- The wire should be *thin* – it will not take a very large load to stretch or break it, so the experiment will be safer and easier to carry out.

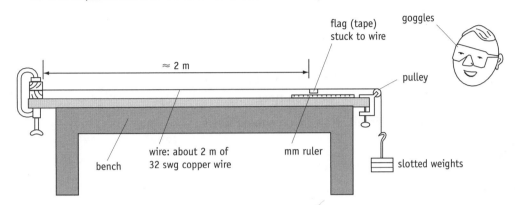

- ***Wear goggles*** – if the wire breaks it will 'whip' into the air and could damage your eyes.

- Measure the diameter with a micrometer and use $A = \pi d^2/4$. (If a micrometer is not available and if you know the gauge of the wire, look up the area in a data book.)

- Measure the original length (from clamp to 'flag') with a metre rule or steel tape measure.

- Add weights one at a time; measure and record the extension each time a weight is added.

- Put all your results in a table.

- Plot force against extension; find $\dfrac{F}{e}$, the gradient of the initial straight-line part of the graph.

- Young modulus $E = \dfrac{Fl}{Ae} = \text{gradient} \times \dfrac{l}{A}$.

There are more accurate methods of measuring the Young modulus, usually with apparatus incorporating a vernier scale to measure the extension.

? Quick check questions

1 A stress of 20 MPa is applied to a wire of Young modulus 10 GPa. What strain is produced? If the wire's initial length was 0.8 m, what extension does the stress produce?

2 The table shows stress–strain data for an iron wire. Use the data to plot a stress–strain graph, deduce the Young modulus for iron, and state whether the wire shows ductile or brittle behaviour.

stress/MPa	0	50	100	150	200	250
strain/10^{-4}	0	2.5	5.0	7.5	10.0	wire breaks

Module 3: end-of-module questions

1 A 12 Ω resistor is connected in a circuit with a 3.0 V battery of negligible internal resistance.

   a Draw a circuit diagram to show this circuit. Add arrows to indicate the directions of conventional current flow, and of electron flow. (2)

   b Calculate the current that flows in the circuit. How much charge flows through the resistor each second? (2)

   c What is the potential difference across the resistor? (1)

   d Calculate the energy transferred in the resistor each second. (2)

2 a Write down an equation linking charge, energy and potential difference. Explain how this equation is used to define potential difference. (2)

   b Use your answer to part **a** to define the volt. (1)

   c The electromotive force (e.m.f.) of a cell can be defined as 'the energy transferred per unit charge when the cell pushes 1 coulomb of charge around a circuit'. Explain how this definition is related to the equation you have stated in part **a**. (2)

3 In the diagram, a resistor R and a thermistor T are connected to a cell of negligible internal resistance.

   a Are the two resistors R and T connected in series or in parallel with one another? (1)

   b A current of 10 mA flows through the resistor R. Calculate the p.d. across it. (2)

   c A current of 4.0 mA flows through T. What is its resistance? (1)

   d The thermistor T is heated so that its resistance decreases. Will the current through it increase, decrease or stay the same? (1)

   e Will the current through R increase, decrease or stay the same? (1)

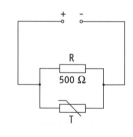

4 In an experiment to investigate the resistance of an alloy of copper, some students use a 0.56 m length of copper alloy wire of diameter 0.40 mm.

   a Calculate the resistance of the wire at room temperature. (Resistivity of the copper alloy at room temperature = 3.4×10^{-7} Ω m.) (3)

The students find that, when a p.d. of 6.0 V is applied across the wire, a current of 400 mA flows through it. Doubling the p.d. to 12.0 V results in a current of 690 mA. They notice that the wire is now hot.

   b **i** Sketch a current–voltage characteristic graph for the copper alloy wire, to show the results that you would expect the students to obtain if they measured current and voltage over a greater range of values. (3)

      ii State and explain whether the copper alloy wire is an ohmic conductor. (2)

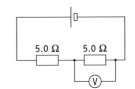

5 A battery of e.m.f. 12.0 V and internal resistance 2.0 Ω is connected as shown to two 5.0 Ω resistors.

    a   Calculate the potential difference measured by the high-impedance voltmeter V. (3)

    b   Calculate the terminal p.d. across the battery. (3)

    c   Explain why the battery's terminal p.d. would increase if the two resistors were replaced with resistors each of value 5 kΩ. (2)

6 In normal use, a 150 W lamp is found to draw a current of 1.5 A from a supply.

    a   Calculate the resistance of the lamp when it is in use. (2)

    b   If the lamp is left switched on for 6 minutes, how many joules of energy are transferred? (2)

7 The figure shows the screen of a cathode ray oscilloscope. A sinusoidal a.c. is connected to the Y input, the Y amplifier is set at 2 V cm^{-1} and the timebase is set at 10 ms cm^{-1}.

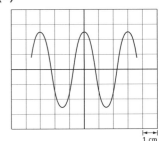

    a   Determine
        i the peak voltage,
        ii the rms voltage,
        iii the time for one cycle,
        iv the frequency. (6)

    b   On a copy of the figure, sketch the trace you would expect to see if the timebase were changed to 5 ms cm^{-1}. (2)

8 An experiment is conducted on a rubber cord of unstretched length 2.0 m. The cord has a rectangular cross-section of dimensions 0.5 mm by 5 mm. The cord is hung vertically and a weight of 100 N attached to its lower end. Its length increases to 2.4 m. Calculate:

    a   the stress in the cord, (3)

    b   the strain, (2)

    c   the Young modulus of the rubber. (2)

9 A vertical wire 1.50 m long and 0.27 mm in diameter is stretched by hanging masses on the end, and the extension measured for each value of the load. The table shows the results.

load/N	0	5.0	10.0	15.0	20.0	22.0	24.0	25.0
extension/mm	0	1.3	2.6	3.9	5.2	5.8	6.7	7.6

    a   i Plot a graph of load (vertical axis) against extension (horizontal axis).
        ii On your graph, indicate the region over which Hooke's law is obeyed. (5)

    b   Use your graph to calculate:
        i the elastic strain energy stored when the wire is under a load of 20.0 N;
        ii a value for the Young modulus of the material of the wire. (6)

10   The figure shows the stress (σ) against strain (ε) graphs to fracture for three different materials A, B and C.

State with a reason which of the materials

    a   is the most ductile, (2)

    b   is brittle, (2)

    c   has the greatest breaking stress, (2)

    d   has the highest value of Young modulus. (2)

Answers to quick check questions

Module 1: Particles, radiation and quantum phenomena

Block 1A

Nuclear structure, pages 2–3

1 $^{28}_{14}\text{Si}$

2 8 of each

3 A and D; B and C

4 more chance of 'direct hit' so more back-scattered

5 5 orders (10^5)

6 10^4

Classifying particles, pages 4–5

1 $+1.6 \times 10^{-19}$ C

2 3.3×10^{-11} J

3 $e^+ + e^- \rightarrow \gamma$ (or $\bar{e} + e \rightarrow \gamma$)

4 hadron, meson

5 three antiquarks: $\bar{u}\,\bar{d}\,\bar{d}$

Interactions between particles, pages 6–7

1 $\bar{u}\,\bar{u}\,\bar{d}$

2 A π^0 is $u\bar{u}$. So an anti-π^0 is $\bar{u}u$, which is the same thing.

3 A neutron decays to form a proton, an electron and an antineutrino.
   An antineutron decays to form an antiproton, a positron and a neutrino.

4

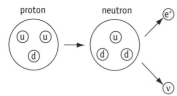

5 Baryon number is not conserved.

Representing forces, pages 8–9

1 intermediate vector boson; photon

2

The down quark emits a W^- vector boson and becomes an up quark. The W^- decays to become an electron and an antineutrino. (This is β^- decay.)

3 $p + \bar{v} \rightarrow n + e^+$

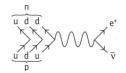

4 $p + e^- \rightarrow n + v$

5 An electron and a positron (antielectron) annihilate to form a photon. The photon decays to form a pair of muons.
   $e^+ + e^- \rightarrow \mu^+ + \mu^-$

Block 1B

Reflection and refraction, pages 10–11

1 55° (both)

2 slows down

3 medium 2

4 26°

Total internal reflection, pages 12–13

1 38.7°

2 56.2°

3 otherwise no total internal reflection

4 Signal would be smeared out.
   Laser light is monochromatic (one wavelength).

Block 1C

The photoelectric effect, pages 14–15

1 – to + (as in any cell)

2 3.6×10^{14} Hz

3 3.9×10^{-20} J

Line spectra, energy levels and ionisation, pages 16–17

1 10.5 eV

2 2.1 eV

3 six

4 2.3×10^{14} Hz

5 15.6 V

Wave–particle duality, pages 18–19

1 slower; larger diameter rings

2 particles

3 6.6×10^{-38} m

4 9.5×10^{-28} kg m s^{-1}

Module 2: Mechanics and molecular kinetic theory

Block 2A

Scalars and vectors, pages 24–25

1 8.6 N at 35.5° to the 5 N force

2 173 N

3 horiz. comp. = 9.1 kN, vert. comp. = 4.2 kN

4 horiz. comp. = 6.6 m s^{-1}, vert. comp. = 4.6 m s^{-1}

Equilibrium, pages 26–27

1 F_1 = 5.77 kN; F_2 = 2.89 kN

2 F_1 = 5.77 kN; F_2 = 2.89 kN

3 2340 N

4 tension = 566 N; pull = 400 N

Turning effects, pages 28–29

1 20 N m; 14.1 N m

2 3 N and 3 N; 15 N m

3 5 N to the right

Block 2B

Velocity and displacement, pages 30–31

1 30 km h^{-1} or 8.3 m s^{-1}

2 200 m s^{-1}

3 40 s

4 Direction changes, so velocity is not constant.

5 9 m s^{-1}

Acceleration, pages 32–33

1 4 m s^{-2}

2 3.0 s

3 −0.6 m s^{-2}; 3750 m

The equations of motion, pages 34–35

1 1.25 m s^{-2}; 250 m s^{-1}

2 19.6 m

3 25 m s^{-1}

4 540 m

5 8.1 m s^{-1}

6 9 s; 189 m

Independence of vertical and horizontal motion, pages 36–37

1 vertical: 7.6 m s^{-1}; horizontal: 16.3 m s^{-1}

2 0.17 s; 0.14 m

3 4.9 m

4 3.5 m

Block 2C

Momentum, pages 38–39

1 mass; kinetic energy

2 1.5×10^6 kg m s^{-1} due west

3 smaller mass × greater velocity = greater mass × smaller velocity

4 Boy has more momentum; girl has more E_k.

Collisions and explosions, pages 40–41

1 10 m s^{-1} in direction of moving car

2 2 m s^{-1} to the right

3 15 m s^{-1}

Newton's laws of motion, pages 42–43

1 zero; constant

2 5000 N

3 two contact forces; two forces on person

4 2 m s^{-1} upwards

5 A

Gravity and motion, pages 44–45

1 44.1 m

2 228 N; 17.1 m

3 Weight is vector; mass is scalar.

4 Velocity decreases to slower, steady value.

Block 2D

Work, energy and power, pages 46–47

1 10 kJ; 4 kJ; 6 kJ

2 1 J = 1 N × 1 m = 1 kg m s^{-2} × 1 m = 1 kg m^2 s^{-2}

3 3×10^6 J

4 force

Energy transfers, pages 48–49

1 vectors; none

2 200 kJ; 1.96 MJ

3 54 m s^{-1}

4 4.9 m s^{-1}

5 Work done against air resistance reduces E_k.

Specific heat capacity and specific latent heat, pages 50–51

1 18 kJ

2 45.5 °C

3 21.4 kW

4 128 kJ

5 2.25×10^6 J kg^{-1}

Block 2E

The equation of state, pages 52–53

1 **a** 361 K; **b** –20 °C
2 3.01×10^{24} particles in both
3 241 K or –32 °C
4 4 litres

Kinetic theory, pages 54–55

1 15.0 m s^{-1}; 15.6 m s^{-1}

2 460 m s^{-1}
3 decreases

Internal energy, pages 56–57

1 Each has half the energy of the original block.
2 400 m s^{-1}
3 6.4×10^{-21} J
4 423 K
5 **a** 1245 J; **b** 311 J; **c** 2.1×10^{-21} J

Module 3: Current electricity and elastic properties of solids

Block 3A

Electric current and voltage, pages 62–63

1 8 mA
2 6000 C
3 9 J
4 42 V
5 60 J
6 4 V

Resistance, pages 64–65

1 5 Ω
2 10 000 V (= 10 kV)
3 2.5 V across each
4 10 Ω
5 18 mA

Ohm's law, pages 66–67

1 $R = 11\ \Omega$ approx.

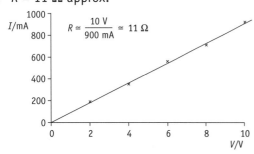

2 6 V; 2 Ω
3 0.09 Ω
4 It increases.

Electrical power, pages 68–69

1 450 MJ (= 450 million J)
2 0.5 W
3 30 W
4 5 A
5 High current – thick cable reduces I so reduces I^2R.

Kirchhoff's laws, pages 70–71

1 5 A

2 14 A
3 160 V
4 6 V; 0.2 A; 1 V, 2 V, 3 V; 6 V = (1 + 2 + 3) V
5 230 J

Internal resistance and potential dividers, pages 72–73

1 3.997 A
2 1.55 V; 20 Ω
3 4 V
4

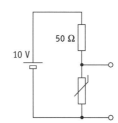

Block 3B

Alternating currents, pages 74–75

1 **a** 44 V; **b** 15.5 V
2 $I_0 = 35$ mA; $V_0 = 0.70$ V
3 50.4 W; 4.2 A

The cathode ray oscilloscope (c.r.o.), pages 76–77

1 **a** 1.5 V; **b** 1.1 V; **c** 5 Hz
2 2.1 μs

Block 3C

Deforming solids, pages 78–79

1 6700 kg
2 0.20 m
3 240 MPa
4 0.33 (no unit)

The Young modulus, pages 80–81

1 0.002; 1.6 mm
2 200 GPa, brittle

Answers to end-of-module questions

Module 1: Particles, radiation and quantum phenomena

Pages 20–22

1 a Both have 10 protons in nucleus and 10 electrons in neutral atom.

  b $^{20}_{10}$Ne has 10 neutrons in nucleus, $^{22}_{10}$Ne has 12 neutrons.

2 a i 6 **ii** 6

  b Nitrogen will not have 6 protons.

3 a straight through

  b

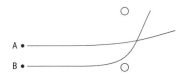

  B must be deflected more than A.

  c Thick film would stop all alpha particles.

  d to prevent deflection by air molecules (alpha particles have short range in air)

4 a electron, muon, neutrino (and their antiparticles)

  b mesons: 2 quarks; baryons: 3 quarks

  c charge $Q = -1$; baryon number = 1

  d charge conserved; baryon number *not* conserved, so reaction never seen

5 a They transmit the force.

  b W^{-1} (an intermediate vector boson)

  c $n + \nu \rightarrow p + e^-$

6 a 1.52

  b 41°

c

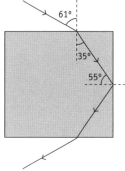

7 a i 1.96×10^8 m s^{-1} **ii** 41°

  b i faster **ii** greater

8 a $n_2/n_1 = \sin \theta_c/\sin 90°$, $\sin 90° = 1$

  b i 67° **ii** 1.43

9 a 4.14 eV

  b minimum energy to remove a conduction electron from the metal

  c 1.94 eV

  d Only the most weakly bound electrons have this E_k. Most of the electrons were more tightly bound within the metal.

10 a i 2.5 eV **ii** 6.0×10^{14} Hz

  b line parallel to that given, and to the right

11 a 5.6×10^{-19} J

  b six

  c 6.0×10^{14} Hz

  d i down arrow from level 3 to level 2

  ii down arrow from level 1 to level 0 (or from 2 to 0 or 3 to 0)

12 a Electrons fired at a graphite film are diffracted.

  b 2.2×10^{-14} m

Module 2: Mechanics and molecular kinetic theory

Pages 58–60

1 a 28.3 kN at 45° to either force

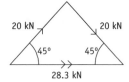

b resistive force = forward force

2 a tension = 3.02 kN

b wind force = 4.16 N

3 a

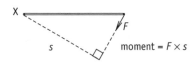

b 4000 N m

c $T \cos 50° \times 3 = 4000$ N m
$T = 4000/\cos 50° \times 3 = 2074$ N

4 a Velocity has direction, speed has only magnitude.

b vectors: force, velocity, acceleration; scalars: distance, kinetic energy, power

5 a acceleration = change in velocity / time

b **i** 1.5 m s^{-2} **ii** 0 m s^{-2}
iii ~ −1.1 m s^{-2} (= slope at X)

6 a $v = u + at$
$t = (20 - 0)/10 = 2$ s to half way, so 4 s total

b

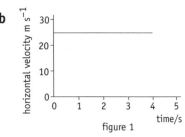

figure 1

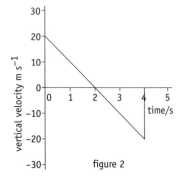

figure 2

c 100 m; 20 m

d 32 m s^{-1}

7 a Motive force is forward friction force of road on tyre.

b reduced friction between tyre and road

c 5 m s^{-2}

d 100 m

8 a Air resistance (drag) decreases as speed increases.

b

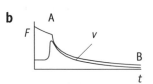

9 a mass × velocity; vector

b 12×10^4 kg m s^{-1}

c When bodies in isolated system interact, total momentum remains constant.

d momentum of child (up) = momentum of earth (down)
before: momentum of child (down) = momentum of earth (up)
after: total momentum zero

10 a No energy is lost.

b momentum before and after = 5 kg × +1 m s^{-1}

11 a $F \times v = 72$ W

b **i** 950 J **ii** 450 J **iii** 1400 J **iv** 19.4 s
v 1.3 m s^{-1}

12 a 3.92 J

b 3.92 J

c E_p reduces as E_k increases.
$E_p + E_k = 3.92$ J

13 a amount of substance (oxygen)

b $pV = RT$

c 124 kPa

14 a N: number of molecules

m: mass of one molecule

$\overline{c^2}$: mean square speed (not r.m.s. speed)

b   **i** 301 K

ii 6.23×10^{-21} J

Module 3: Current electricity and elastic properties of solids

Pages 82–83

1 a

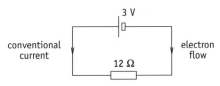

b 0.25 A; 0.25 C

c 3 V

d 0.75 J

2 a $V = W/Q$; p.d. = energy transferred per coulomb

b $1 \text{ V} = 1 \text{ J C}^{-1}$

c $W = 1$ J, $Q = 1$ C

3 a in parallel

b 5 V

c 1250 Ω

d increase

e stay the same

4 a 1.51 Ω

b   **i**

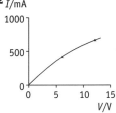

ii The I–V graph is not a straight line through the origin, so the copper alloy wire is non-ohmic.

5 a 5.0 V

b 10.0 V

c smaller I, so less p.d. across internal resistance

6 a 66.7 Ω

b 54 kJ

7 a **i** 5.0 V **ii** 3.6 V **iii** 30 ms **iv** 33.3 Hz

b Trace has same amplitude but one wave now covers 6 cm instead of three.

8 a 40 MPa

b 0.2

c 200 MPa

9 a

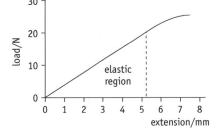

b **i** 52 mJ **ii** 101 GPa

10 a C (greatest strain)

b B (no plastic region)

c A (highest curve)

d B (steepest slope)

Index

0161 872 5050

56

Heinemann Educational Publishers
Halley Court, Jordan Hill, Oxford OX2 8EJ
Part of Harcourt Education
Heinemann is the registered trademark of Harcourt Education Limited

© Harvey Cole and David Sang, 2001

First published 2001

10-Digit ISBN 0 435583 36 0
13-Digit ISBN 978 0 435583 36 1

05
10 9 8 7 6 5

Development Editor Paddy Gannon

Edited by Patrick Bonham

Designed and typeset by Saxon Graphics Ltd, Derby

Index compiled by Paul Nash

Printed and bound in Great Britain by Thomson Litho Ltd, Glasgow

Tel. 01865 888058, www.heinemann.co.uk